Algebra through practice

D0783226

Book 4: Linear algebra

Algebra through practice

A collection of problems in algebra with solutions

Book 4
Linear algebra

T. S. BLYTH o E. F. ROBERTSON
University of St Andrews

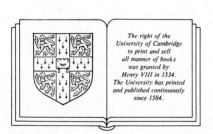

The right of the
University of Cambridge
to print and sell
all manner of books
was granted by
Henry VIII in 1534.
The University has printed
and published continuously
since 1584.

CAMBRIDGE UNIVERSITY PRESS

Cambridge

London New York New Rochelle

Melbourne Sydney

Published by the Press Syndicate of the University of Cambridge
The Pitt Building, Trumpington Street, Cambridge CB2 IRP
32 East 57th Street, New York, NY 10022, USA
10 Stamford Road, Oakleigh, Melbourne 3166, Australia

First published 1985

Printed in Great Britain at the University Press, Cambridge

Library of Congress catalogue card number: 83–24013

British Library cataloguing in publication data

Blyth, T. S.
   Algebra through practice: a collection of problems in algebra with solutions.
   Bk. 4: Linear algebra
   1. Algebra—Problems, exercises, etc.
   I. Title   II. Robertson, E. F.
   512′.0076   QA157

ISBN 0 521 27289 0

Contents

Preface

The aim of this series of problem-solvers is to provide a selection of worked examples in algebra designed to supplement undergraduate algebra courses. We have attempted, mainly with the average student in mind, to produce a varied selection of exercises while incorporating a few of a more challenging nature. Although complete solutions are included, it is intended that these should be consulted by readers only after they have attempted the questions. In this way, it is hoped that the student will gain confidence in his or her approach to the art of problem-solving which, after all, is what mathematics is all about.

The problems, although arranged in chapters, have not been 'graded' within each chapter so that, if readers cannot do problem n this should not discourage them from attempting problem $n+1$. A great many of the ideas involved in these problems have been used in examination papers of one sort or another. Some test papers (without solutions) are included at the end of each book; these contain questions based on the topics covered.

TSB, EFR
St Andrews

Background reference material

Courses on abstract algebra can be very different in style and content. Likewise, textbooks recommended for these courses can vary enormously, not only in notation and exposition but also in their level of sophistication. Here is a list of some major texts that are widely used and to which the reader may refer for background material. The subject matter of these texts covers all six of the present volumes, and in some cases a great deal more. For the convenience of the reader there is given overleaf an indication of which parts of which of these texts are most relevant to the appropriate sections of this volume.

[1] I. T. Adamson, *Introduction to Field Theory*, Cambridge University Press, 1982.

[2] F. Ayres, Jr, *Modern Algebra*, Schaum's Outline Series, McGraw-Hill, 1965.

[3] D. Burton, *A first course in rings and ideals*, Addison-Wesley, 1970.

[4] P. M. Cohn, *Algebra* Vol. I, Wiley, 1982.

[5] D. T. Finkbeiner II, *Introduction to Matrices and Linear Transformations*, Freeman, 1978.

[6] R. Godement, *Algebra*, Kershaw, 1983.

[7] J. A. Green, *Sets and Groups*, Routledge and Kegan Paul, 1965.

[8] I. N. Herstein, *Topics in Algebra*, Wiley, 1977.

[9] K. Hoffman and R. Kunze, *Linear Algebra*, Prentice Hall, 1971.

[10] S. Lang, *Introduction to Linear Algebra*, Addison-Wesley, 1970.

[11] S. Lipschutz, *Linear Algebra*, Schaum's Outline Series, McGraw-Hill, 1974.

[12] I. D. Macdonald, *The Theory of Groups*, Oxford University Press, 1968.

[13] S. MacLane and G. Birkhoff, *Algebra*, Macmillan, 1968.

[14] N. H. McCoy, *Introduction to Modern Algebra*, Allyn and Bacon, 1975.

[15] J. J. Rotman, *The Theory of Groups: An Introduction*, Allyn and Bacon, 1973.

[16] I. Stewart, *Galois Theory*, Chapman and Hall, 1975.

[17] I. Stewart and D. Tall, *The Foundations of Mathematics*, Oxford University Press, 1977.

References useful for Book 4

1: Direct sums and Jordan forms   [**4**, Sections 11.1–11.4], [**5**, Chapter 7], [**8**, Sections 6.1–6.6], [**9**, Chapters 6, 7], [**11**, Chapter 10].

2: Duality and normal transformations   [**4**, Chapter 8, Section 11.4], [**5**, Chapter 9], [**8**, Sections 4.3, 6.8, 6.10], [**9**, Chapters 8, 9], [**11**, Chapters 11, 12].

In [**4**] and [**6**] some ring theory is assumed, and some elementary results are proved for modules. In [**5**] the author uses 'characteristic value' where we use 'eigenvalue'.

1: Direct sums and Jordan forms

In this chapter we take as a central theme the notion of the *direct sum* $A \oplus B$ of subspaces A, B of a vector space V. Recall that $V = A \oplus B$ if and only if every $x \in V$ can be expressed uniquely in the form $a + b$ where $a \in A$ and $b \in B$; equivalently, if $V = A + B$ and $A \cap B = \{0\}$. For every subspace A of V there is a subspace B of V such that $V = A \oplus B$. In the case where V is of finite dimension, this is easily seen; take a basis $\{v_1, \ldots, v_k\}$ of A, extend it to a basis $\{v_1, \ldots, v_n\}$ of V, then note that $\{v_{k+1}, \ldots, v_n\}$ spans a subspace B such that $V = A \oplus B$.

If $f : V \to V$ is a linear transformation then a subspace W of V is said to be *f-invariant* if f maps W into itself. If W is f-invariant then there is an ordered basis of V with respect to which the matrix of V is of the form

$$\begin{bmatrix} M & N \\ 0 & X \end{bmatrix}$$

where M is of size $\dim W \times \dim W$.

If $f : V \to V$ is such that $f \circ f = f$ then f is called a *projection*. For such a linear transformation we have $V = \operatorname{Im} f \oplus \operatorname{Ker} f$ where the subspace $\operatorname{Im} f$ is f-invariant (and the subspace $\operatorname{Ker} f$ is trivially so). A vector space V is the direct sum of subspaces $W_1, \ldots, W_k$ if and only if there are non-zero projections $p_1, \ldots, p_k : V \to V$ such that

$$\sum_{i=1}^{k} p_i = \operatorname{id}_V \qquad \text{and} \qquad p_i \circ p_j = 0 \quad \text{for} \quad i \neq j.$$

In this case $W_i = \operatorname{Im} p_i$ for each i, and relative to given ordered bases of

$W_1, \ldots, W_k$ the matrix of f is of the diagonal block form

$$\begin{bmatrix} M_1 & & & \\ & M_2 & & \\ & & \ddots & \\ & & & M_k \end{bmatrix}.$$

Of particular importance is the situation where each M_i is of the form

$$M_i = \begin{bmatrix} \lambda & 1 & 0 & \ldots & 0 & 0 \\ 0 & \lambda & 1 & \ldots & 0 & 0 \\ 0 & 0 & \lambda & \ldots & 0 & 0 \\ \vdots & \vdots & \vdots & \ddots & \vdots & \vdots \\ 0 & 0 & 0 & \ldots & \lambda & 1 \\ 0 & 0 & 0 & \ldots & 0 & \lambda \end{bmatrix}$$

in which case the diagonal block matrix is called a *Jordan matrix*.

The Cayley–Hamilton theorem says that a linear transformation f is a zero of its characteristic polynomial. The minimum polynomial of f is the monic polynomial of least degree of which f is a zero. When the minimum polynomial of f factorises into a product of linear polynomials then there is a basis of V with respect to which the matrix of f is a Jordan matrix. This matrix is unique (up to the sequence of the diagonal blocks), the diagonal entries λ above are the eigenvalues of f, and the number of M_i associated with a given λ is the geometric multiplicity of λ. The corresponding basis is called a *Jordan basis*.

We mention here that, for space considerations in the solutions, we shall often write an eigenvector

$$\begin{bmatrix} x_1 \\ x_2 \\ \vdots \\ x_n \end{bmatrix}$$

as $[x_1, x_2, \ldots, x_n]$.

1.1   Which of the following statements are true? For those that are false, give a counter-example.

    (i) If $\{a_1, a_2, a_3\}$ is a basis for $\mathbb{R}^3$ and b is a non-zero vector in $\mathbb{R}$ then $\{b + a_1, a_2, a_3\}$ is also a basis for $\mathbb{R}^3$.

(ii) If A is a finite set of linearly independent vectors then the dimension of the subspace spanned by A is equal to the number of vectors in A.

(iii) The subspace $\{(x, x, x) \mid x \in \mathbb{R}\}$ of $\mathbb{R}^3$ has dimension 3.

(iv) If A is a linearly dependent set of vectors in $\mathbb{R}^n$ then there are more than n vectors in A.

(v) If A is a linearly dependent subset of $\mathbb{R}^n$ then the dimension of the subspace spanned by A is strictly less than the number of vectors in A.

(vi) If A is a subset of $\mathbb{R}^n$ and the subspace spanned by A is $\mathbb{R}^n$ itself then A contains exactly n vectors.

(vii) If A and B are subspaces of $\mathbb{R}^n$ then we can find a basis of $\mathbb{R}^n$ which contains a basis of A and a basis of B.

(viii) An n–dimensional vector space contains only finitely many subspaces.

(ix) If A is an $n \times n$ matrix over $\mathbb{Q}$ with $A^3 = I$ then A is non-singular.

(x) If A is an $n \times n$ matrix over $\mathbb{C}$ with $A^3 = I$ then A is non-singular.

(xi) An isomorphism between two vector spaces can always be represented by a square singular matrix.

(xii) Any two n–dimensional vector spaces are isomorphic.

(xiii) If A is an $n \times n$ matrix such that $A^2 = I$ then $A = I$.

(xiv) If A, B and C are non-zero matrices such that $AC = BC$ then $A = B$.

(xv) The identity map on $\mathbb{R}^n$ is represented by the identity matrix with respect to any basis of $\mathbb{R}^n$.

(xvi) Given any two bases of $\mathbb{R}^n$ there is an isomorphism from $\mathbb{R}^n$ to itself that maps one basis onto the other.

(xvii) If A and B represent linear transformations $f, g : \mathbb{R}^n \to \mathbb{R}^n$ with respect to the same basis then there is a non-singular matrix P such that $P^{-1}AP = B$.

(xviii) There is a bijection between the set of linear transformations from $\mathbb{R}^n$ to itself and the set of $n \times n$ matrices over $\mathbb{R}$.

(xix) The map $t : \mathbb{R}^2 \to \mathbb{R}^2$ given by $t(x, y) = (y, x+y)$ can be represented by the matrix

$$\begin{bmatrix} 1 & 2 \\ 1 & 2 \end{bmatrix}$$

with respect to some basis of $\mathbb{R}^2$.

(xx) There is a non-singular matrix P such that $P^{-1}AP$ is diagonal for any non-singular matrix A.

1.2   Let $t_1, t_2, t_3, t_4 \in \mathcal{L}(\mathbb{R}^3, \mathbb{R}^3)$ be given by

$$t_1(a, b, c) = (a + b, b + c, c + a);$$
$$t_2(a, b, c) = (a - b, b - c, 0);$$
$$t_3(a, b, c) = (-b, a, c);$$
$$t_4(a, b, c) = (a, b, b).$$

Find $\mathrm{Ker}\, t_i$ and $\mathrm{Im}\, t_i$ for $i = 1, 2, 3, 4$. Is it true that $\mathbb{R}^3 = \mathrm{Ker}\, t_i \oplus \mathrm{Im}\, t_i$ for any of $i = 1, 2, 3, 4$?

Is $\mathrm{Im}\, t_2$ t_3–invariant? Is $\mathrm{Ker}\, t_2$ t_3–invariant?

Find $t_3 \circ t_4$ and $t_4 \circ t_3$. Compute the images and kernels of these composites.

1.3   Let V be a vector space of dimension 3 over a field F and let $t \in \mathcal{L}(V, V)$ be represented by the matrix

$$\begin{bmatrix} 3 & -1 & 1 \\ -1 & 5 & -1 \\ 1 & -1 & 3 \end{bmatrix}$$

with respect to some basis of V. Find $\dim \mathrm{Ker}\, t$ and $\dim \mathrm{Im}\, t$ when

(i) $F = \mathbb{R}$;
(ii) $F = \mathbb{Z}_2$;
(iii) $F = \mathbb{Z}_3$.

Is $V = \mathrm{Ker}\, t \oplus \mathrm{Im}\, t$ in any of cases (i), (ii) or (iii)?

1.4   Let V be a finite-dimensional vector space and let $s, t \in \mathcal{L}(V, V)$ be such that $s \circ t = \mathrm{id}_V$. Prove that $t \circ s = \mathrm{id}_V$. Prove also that a subspace of V is t–invariant if and only if it is s–invariant. Are these results true when V is infinite-dimensional?

1.5   Let V_n be the vector space of polynomials of degree less than n over the field $\mathbb{R}$. If $D \in \mathcal{L}(V_n, V_n)$ is the differentiation map, find $\mathrm{Im}\, D$ and $\mathrm{Ker}\, D$. Prove that $\mathrm{Im}\, D \simeq V_{n-1}$ and that $\mathrm{Ker}\, D \simeq \mathbb{R}$. Is it true that

$$V_n = \mathrm{Im}\, D \oplus \mathrm{Ker}\, D \ ?$$

Do the same results hold if the ground field $\mathbb{R}$ is replaced by the field $\mathbb{Z}_2$?

1.6 Let V be a finite-dimensional vector space and let $t \in \mathcal{L}(V, V)$. Establish the chains

$$V \supseteq \operatorname{Im} t \supseteq \operatorname{Im} t^2 \supseteq \ldots \supseteq \operatorname{Im} t^n \supseteq \operatorname{Im} t^{n+1} \supseteq \ldots;$$
$$\{0\} \subseteq \operatorname{Ker} t \subseteq \operatorname{Ker} t^2 \subseteq \ldots \subseteq \operatorname{Ker} t^n \subseteq \operatorname{Ker} t^{n+1} \subseteq \ldots.$$

Show that there is a positive integer p such that $\operatorname{Im} t^p = \operatorname{Im} t^{p+1}$ and deduce that

$$(\forall k \geq 1) \qquad \operatorname{Im} t^p = \operatorname{Im} t^{p+k} \quad \text{and} \quad \operatorname{Ker} t^p = \operatorname{Ker} t^{p+k}.$$

Show also that

$$V = \operatorname{Im} t^p \oplus \operatorname{Ker} t^p$$

and that the subspaces $\operatorname{Im} t^p$ and $\operatorname{Ker} t^p$ are t–invariant.

1.7 Let V be a vector space of dimension n over a field F and let $f : V \to V$ be a non-zero linear transformation such that $f \circ f = 0$. Show that if $\operatorname{Im} f$ is of dimension r then $2r \leq n$. Suppose now that W is a subspace of V such that $V = \operatorname{Ker} f \oplus W$. Show that W is of dimension r and that if $\{w_1, \ldots, w_r\}$ is a basis of W then $\{f(w_1), \ldots, f(w_r)\}$ is a linearly independent subset of $\operatorname{Ker} f$. Deduce that $n - 2r$ elements $x_1, \ldots, x_{n-2r}$ can be chosen in $\operatorname{Ker} f$ such that

$$\{w_1, \ldots, w_r, f(w_1), \ldots, f(w_r), x_1, \ldots, x_{n-2r}\}$$

is a basis of V.

Hence show that a non-zero $n \times n$ matrix A over F is such that $A^2 = 0$ if and only if A is similar to a matrix of the form

$$\begin{bmatrix} 0_r & 0 \\ I_r & 0 \\ 0 & 0 \end{bmatrix}.$$

1.8 Let V be a vector space of dimension 4 over $\mathbb{R}$. Let a basis of V be $B = \{b_1, b_2, b_3, b_4\}$. Writing each $x \in V$ as $x = \sum_{i=1}^{4} x_i b_i$, let

$$V_1 = \{x \in V \mid x_3 = x_2 \text{ and } x_4 = x_1\},$$
$$V_2 = \{x \in V \mid x_3 = -x_2 \text{ and } x_4 = -x_1\}.$$

Show that
(1) V_1 and V_2 are subspaces of V;

(2) $\{b_1 + b_4, b_2 + b_3\}$ is a basis of V_1 and $\{b_1 - b_4, b_2 - b_3\}$ is a basis of V_2;

(3) $V = V_1 \oplus V_2$;

(4) with respect to the basis B and the basis

$$C = \{b_1 + b_4, b_2 + b_3, b_2 - b_3, b_1 - b_4\}$$

the matrix of id_V is

$$\begin{bmatrix} \frac{1}{2} & 0 & 0 & \frac{1}{2} \\ 0 & \frac{1}{2} & \frac{1}{2} & 0 \\ 0 & \frac{1}{2} & -\frac{1}{2} & 0 \\ \frac{1}{2} & 0 & 0 & -\frac{1}{2} \end{bmatrix}.$$

A 4×4 matrix M over $\mathbb{R}$ is said to be *centro-symmetric* if

$$m_{ij} = m_{5-i,5-j}$$

for all i, j. If M is centro-symmetric, show that M is similar to a matrix of the form

$$\begin{bmatrix} \alpha & \beta & 0 & 0 \\ \gamma & \delta & 0 & 0 \\ 0 & 0 & \epsilon & \varsigma \\ 0 & 0 & \eta & \vartheta \end{bmatrix}.$$

1.9   Let V be a vector space of dimension n over a field F. Suppose first that F is not of characteristic 2 (i.e. that $1_F + 1_F \neq 0_F$). If $f : V \to V$ is a linear transformation such that $f \circ f = \mathrm{id}_V$ prove that

$$V = \mathrm{Im}(\mathrm{id}_V + f) \oplus \mathrm{Im}(\mathrm{id}_V - f).$$

Deduce that an $n \times n$ matrix A over F is such that $A^2 = I_n$ if and only if A is similar to a matrix of the form

$$\begin{bmatrix} I_p & 0 \\ 0 & -I_{n-p} \end{bmatrix}.$$

Suppose now that F is of characteristic 2 and that $f \circ f = \mathrm{id}_V$. If $g = \mathrm{id}_V + f$ show that

$$x \in \mathrm{Ker}\, g \iff x = f(x),$$

and that $g \circ g = 0$. Deduce that an $n \times n$ matrix A over F is such that $A^2 = I_n$ if and only if A is similar to a matrix of the form

$$
\begin{bmatrix}
I_{n-2p} & & & & & & \\
& 1 & 1 & & & & \\
& 0 & 1 & & & & \\
& & & 1 & 1 & & \\
& & & 0 & 1 & & \\
& & & & & \ddots & \\
& & & & & 1 & 1 \\
& & & & & 0 & 1
\end{bmatrix}.
$$

[*Hint.* Observe that $\operatorname{Im} g \subseteq \operatorname{Ker} g$. Let $\{g(c_1), \ldots, g(c_p)\}$ be a basis of $\operatorname{Im} g$ and extend this to a basis $\{b_1, \ldots, b_{n-2p}, g(c_1), \ldots, g(c_p)\}$ of $\operatorname{Ker} g$. Show that

$$\{b_1, \ldots, b_{n-2p}, g(c_1), c_1, \ldots, g(c_p), c_p\}$$

is a basis of V.]

1.10 Let V be a finite-dimensional vector space and let $t \in \mathcal{L}(V, V)$ be such that $t \neq \operatorname{id}_V$ and $t \neq 0$. Is it possible to have $\operatorname{Im} t \cap \operatorname{Ker} t \neq \{0\}$? Is it possible to have $\operatorname{Im} t = \operatorname{Ker} t$? Is it possible to have $\operatorname{Im} t \subset \operatorname{Ker} t$? Is it possible to have $\operatorname{Ker} t \subset \operatorname{Im} t$? Which of these are possible if t is a projection?

1.11 Is it possible to have projections $e, f \in \mathcal{L}(V, V)$ with $\operatorname{Ker} e = \operatorname{Ker} f$ and $\operatorname{Im} e \neq \operatorname{Im} f$? Is it possible to have $\operatorname{Im} e = \operatorname{Im} f$ and $\operatorname{Ker} e \neq \operatorname{Ker} f$? Is it possible to have projections e, f with $e \circ f = 0$ but $f \circ e \neq 0$?

1.12 Let V be a vector space over a field of characteristic not equal to 2. Let $e_1, e_2 \in \mathcal{L}(V, V)$ be projections. Prove that $e_1 + e_2$ is a projection if and only if $e_1 \circ e_2 = e_2 \circ e_1 = 0$.

If $e_1 + e_2$ is a projection, find $\operatorname{Im}(e_1 + e_2)$ and $\operatorname{Ker}(e_1 + e_2)$ in terms of the images and kernels of e_1, e_2.

1.13 Let V be the subspace of $\mathbb{R}^3$ given by

$$V = \{(a, a, 0) \mid a \in \mathbb{R}\}.$$

Find a subspace U of $\mathbb{R}^3$ such that $\mathbb{R}^3 = V \oplus U$. Is U unique? Find a projection $e \in \mathcal{L}(\mathbb{R}^3, \mathbb{R}^3)$ such that $\operatorname{Im} e = V$ and $\operatorname{Ker} e = U$. Find also a projection $f \in \mathcal{L}(\mathbb{R}^3, \mathbb{R}^3)$ such that $\operatorname{Im} f = U$ and $\operatorname{Ker} f = V$.

1.14 If V is a finite-dimensional vector space over a field F and $e, f \in \mathcal{L}(V, V)$ are projections prove that $\operatorname{Im} e = \operatorname{Im} f$ if and only if $e \circ f = f$ and $f \circ e = e$.

Suppose that $e_1, \ldots, e_k \in \mathcal{L}(V, V)$ are projections with
$$\operatorname{Im} e_1 = \operatorname{Im} e_2 = \cdots = \operatorname{Im} e_k.$$
Let $\lambda_1, \lambda_2, \ldots, \lambda_k \in F$ be such that $\sum_{i=1}^{k} \lambda_i = 1$. Prove that
$$e = \lambda_1 e_1 + \lambda_2 e_2 + \cdots + \lambda_k e_k$$
is a projection with $\operatorname{Im} e = \operatorname{Im} e_i$.

Is it necessarily true that if $f_1, \ldots, f_k \in \mathcal{L}(V, V)$ are projections and $\sum_{i=1}^{k} \lambda_i = 1$ then $\sum_{i=1}^{k} \lambda_i f_i$ is also a projection?

1.15 A *net* over the interval $[0, 1]$ of $\mathbb{R}$ is a finite sequence $(a_i)_{0 \leq i \leq n+1}$ such that
$$0 = a_0 < a_1 < \cdots < a_n < a_{n+1} = 1.$$

A *step function* on $[0, 1[$ is a mapping $f : [0, 1[\to \mathbb{R}$ for which there exists a net $(a_i)_{0 \leq i \leq n+1}$ over $[0, 1]$ and a finite sequence $(b_i)_{0 \leq i \leq n}$ of elements of $\mathbb{R}$ such that
$$(\forall x \in [a_i, a_{i+1}[) \qquad f(x) = b_i.$$

Show that the set E of step functions on $[0, 1[$ is a vector space over $\mathbb{R}$ and that a basis of E is the set $\{e_k \mid k \in [0, 1[\}$ of functions $e_k : [0, 1[\to \mathbb{R}$ given by
$$e_k(x) = \begin{cases} 0 & \text{if } 0 \leq x < k; \\ 1 & \text{if } k \leq x < 1. \end{cases}$$

A *piecewise linear function* on $[0, 1[$ is a mapping $f : [0, 1[\to \mathbb{R}$ for which there exists a net $(a_i)_{0 \leq i \leq n+1}$ and sequences $(b_i)_{0 \leq i \leq n}, (c_i)_{0 \leq i \leq n}$ of elements of $\mathbb{R}$ such that
$$(\forall x \in [a_i, a_{i+1}[) \qquad f(x) = b_i x + c_i.$$

Let F be the set of piecewise linear functions on $[0, 1[$ and let G be the subset of F consisting of the piecewise linear functions g that are continuous with $g(0) = 0$. Show that F, G are vector spaces over $\mathbb{R}$ and that $F = E \oplus G$.

Show that a basis of G is the set $\{g_k \mid k \in [0, 1[\}$ of functions given by
$$g_k(x) = \begin{cases} 0 & \text{if } 0 \leq x < k; \\ x - k & \text{if } k \leq x < 1. \end{cases}$$

Finally, show that the assignment
$$f \longmapsto I(f) = \int_0^x f(t)\, dt$$
describes an isomorphism from E to G.

1: Direct sums and Jordan forms

1.16 Let V be a vector space over a field F and let $t \in \mathcal{L}(V, V)$. Let λ_1 and λ_2 be distinct eigenvalues of t with associated eigenvectors v_1 and v_2. Is it possible for $\lambda_1 + \lambda_2$ to be an eigenvalue of t? What about $\lambda_1 \lambda_2$?

1.17 Let $t \in \mathcal{L}(\mathbb{C}^2, \mathbb{C}^2)$ be given by

$$t(a, b) = (a + 2b, b - a).$$

Find the eigenvalues of t and show that there is a basis of $\mathbb{C}^2$ consisting of eigenvectors of t. Find such a basis, and the matrix of t with respect to this basis.

1.18 Suppose that $t \in \mathcal{L}(V, V)$ has zero as an eigenvalue. Prove that t is not invertible. Is it true that t is invertible if and only if all the eigenvalues of t are non-zero? If t is invertible, how are the eigenvalues of t related to those of t^{-1}?

1.19 Let V be a vector space of finite dimension over $\mathbb{Q}$ and let $t \in \mathcal{L}(V, V)$ be such that $t^m = 0$ for some $m > 0$. Prove that all the eigenvalues of t are zero. Deduce that if $t \neq 0$ then t is not diagonalisable.

1.20 Consider $t \in \mathcal{L}(\mathbb{R}^2, \mathbb{R}^2)$ given by

$$t(a, b) = (a + 4b, \tfrac{1}{2}a - b).$$

Find the minimum polynomial of t.

1.21 Let F be a field and let $F_{n+1}[X]$ be the vector space of polynomials of degree less than or equal to n over F. Define $t : F_{n+1}[X] \to F_{n+1}[X]$ by $t(f(X)) = f(X + 1)$. Show that t is linear.
 Find the matrix of t relative to the basis $\{1, X, \dots, X^n\}$ of $F_{n+1}[X]$. Find also the eigenvalues of t. If $g(X) = (X - 1)^{n+1}$ show that $g(t) = 0$. Hence find the minimum polynomial of t.

1.22 If V is a finite-dimensional vector space and $t \in \mathcal{L}(V, V)$ is such that $t^2 = \mathrm{id}_V$ prove that the sum of the eigenvalues of t is an integer.

1.23 For each of the following real matrices, determine

(i) the eigenvalues;
(ii) the geometric multiplicity of each eigenvalue;
(iii) whether it is diagonalisable.

9

For those matrices A that are diagonalisable, find an invertible matrix P such that $P^{-1}AP$ is diagonal.

$$(a) \begin{bmatrix} 3 & -1 \\ -1 & 3 \end{bmatrix}, \quad (b) \begin{bmatrix} 3 & -1 & 1 \\ -1 & 5 & -1 \\ 1 & -1 & 3 \end{bmatrix}, \quad (c) \begin{bmatrix} 7 & -1 & 2 \\ -1 & 7 & 2 \\ -2 & 2 & 10 \end{bmatrix},$$

$$(d) \begin{bmatrix} 2 & 1 & -1 \\ 0 & 2 & 1 \\ 0 & 0 & 1 \end{bmatrix}, \quad (e) \begin{bmatrix} 1 & 0 & 1 \\ 0 & 2 & 1 \\ -1 & 0 & 3 \end{bmatrix}.$$

1.24 Consider the sequence described by

$$\frac{1}{1}, \frac{3}{2}, \frac{7}{5}, \ldots, \frac{a_n}{b_n}, \ldots$$

where $a_{n+1} = a_n + 2b_n$ and $b_{n+1} = a_n + b_n$.
Find a matrix A such that

$$\begin{bmatrix} a_{n+1} \\ b_{n+1} \end{bmatrix} = A \begin{bmatrix} a_n \\ b_n \end{bmatrix}.$$

By diagonalising A, obtain explicit formulae for a_n and b_n and hence show that

$$\lim_{n \to \infty} \frac{a_n}{b_n} = \sqrt{2}.$$

1.25 Let t be a singular transformation on a real vector space V. Let $f(X)$ and $g(X)$ be real polynomials whose highest common factor is 1. Let $a = f(t)$ and $b = g(t)$.

Prove that every eigenvector of $a \circ b$ that is associated with the eigenvalue 0 is the sum of an eigenvector of a associated with the eigenvalue 0 and an eigenvector of b associated with the eigenvalue 0.

1.26 Suppose that $s, t \in \mathcal{L}(V, V)$ each have $n = \dim V$ distinct eigenvalues and that $s \circ t = t \circ s$. Prove that, for every λ in the ground field F,

$$C_\lambda = \{v \in V \mid t(v) = \lambda v\}$$

is a subspace of V. Show that C_λ is s–invariant and that, when λ_i is an eigenvalue of t, the subspace C_{λ_i} has dimension 1.

Hence show that the matrix of s with respect to the basis of eigenvectors of t is diagonal.

1.27 Let u be a non-zero vector in an n-dimensional vector space V over a field F, let $t \in \mathcal{L}(V,V)$, and let U be the subspace spanned by

$$\{u, t(u), t^2(u), \ldots, t^{n-1}(u)\}.$$

Show that there is a greatest integer r such that the set

$$\{u, t(u), t^2(u), \ldots, t^{r-1}(u)\}$$

is linearly independent and deduce that this set is a basis for U. Show also that U is t-invariant.

Show that there is a non-zero monic polynomial $f(X) \in F[X]$ of degree r such that $[f(t)](u) = 0$. If $t_U : U \to U$ is the linear transformation induced by t, show that its minimum polynomial is $f(X)$.

In the case where $u = (1,1,0) \in \mathbb{R}^3$ and t is given by

$$t(x,y,z) = (x+y, x-y, z),$$

find the minimum polynomial of t_U.

1.28 Let r, s, t be non-zero linear transformations on a finite-dimensional vector space V such that $r \circ t \circ r = 0$. Let $p = r \circ s$ and $q = r \circ (s+t)$ and suppose that the minimum polynomials of p, q are $p(X), q(X)$ respectively. Prove that (with composites written as products)

(1) $q^n = p^{n-1}q$ and $p^n = q^{n-1}p$ for all $n \geq 1$;

(2) $p(X)$ and $q(X)$ are divisible by X;

(3) q satisfies $Xp(X) = 0$, and p satisfies $Xq(X) = 0$.

Deduce that one of the following holds :

(i) $p(X) = q(X)$;

(ii) $p(X) = Xq(X)$;

(iii) $q(X) = Xp(X)$.

1.29 A 3×3 complex matrix M is said to be *magic* if every row sum, every column sum, and both diagonal sums are equal to some $\vartheta \in \mathbb{C}$.

If M is magic, prove that $\vartheta = 3m_{22}$. Deduce that, given $\alpha, \beta, \gamma \in \mathbb{C}$, there is a unique magic matrix $M(\alpha, \beta, \gamma)$ such that

$$m_{22} = \alpha, \quad m_{11} = \alpha + \beta, \quad m_{31} = \alpha + \gamma.$$

Show that $\{M(\alpha, \beta, \gamma) \mid \alpha, \beta, \gamma \in \mathbb{C}\}$ is a subspace of $\text{Mat}_{3\times3}(\mathbb{C})$ and that

$$B = \{M(1,0,0), M(0,1,0), M(0,0,1)\}$$

is a basis of this subspace.

If $f : \mathbb{C}^3 \to \mathbb{C}^3$ represents $M(\alpha, \beta, \gamma)$ relative to the canonical basis $\{e_1, e_2, e_3\}$, show that $e_1 + e_2 + e_3$ is an eigenvector of f. Determine the matrix of f relative to the basis $\{e_1 + e_2 + e_3, e_2, e_3\}$. Hence find the eigenvalues of $M(\alpha, \beta, \gamma)$.

1.30 Let V be a vector space of dimension n over a field F. A linear transformation $f : V \to V$ (respectively, an $n \times n$ matrix A over F) is said to be *nilpotent of index p* if there is an integer $p > 1$ such that $f^{p-1} \neq 0$ and $f^p = 0$ (respectively, $A^{p-1} \neq 0$ and $A^p = 0$).

Show that if f is nilpotent of index p and $x \in V \backslash \{0\}$ is such that $f^{p-1}(x) \neq 0$ then
$$\{x, f(x), \ldots, f^{p-1}(x)\}$$
is a linearly independent subset of V. Hence show that f is nilpotent of index n if and only if there is an ordered basis of V with respect to which the matrix of f is
$$\begin{bmatrix} 0 & 0 \\ I_{n-1} & 0 \end{bmatrix}.$$

Deduce that an $n \times n$ matrix A over F is nilpotent of index n if and only if A is similar to this matrix.

1.31 Let V be a finite-dimensional vector space over $\mathbb{R}$ and let $f : V \to V$ be a linear transformation such that $f \circ f = -\operatorname{id}_V$. Extend the external law $\mathbb{R} \times V \to V$ to an external law $\mathbb{C} \times V \to V$ by defining, for all $x \in V$ and all $\alpha + i\beta \in \mathbb{C}$,

$$(\alpha + i\beta)x = \alpha x - \beta f(x).$$

Show that in this way V becomes a vector space over $\mathbb{C}$. Use the identity

$$\sum_{t=1}^{r} (\lambda_t - i\mu_t)v_t = \sum_{t=1}^{r} \lambda_t v_t + \sum_{t=1}^{r} \mu_t f(v_t)$$

to show that if $\{v_1, \ldots, v_r\}$ is a linearly independent subset of the $\mathbb{C}$-vector space V then $\{v_1, \ldots, v_r, f(v_1), \ldots, f(v_r)\}$ is a linearly independent subset of the $\mathbb{R}$-vector space V. Deduce that the dimension of V as a vector space over $\mathbb{C}$ is finite, n say, and that $\dim_{\mathbb{R}} V = 2n$.

Hence show that a $2n \times 2n$ matrix A over $\mathbb{R}$ is such that $A^2 = -I_{2n}$ if and only if A is similar to the matrix

$$\begin{bmatrix} 0 & -I_n \\ I_n & 0 \end{bmatrix}.$$

1.32 Let A be a real skew-symmetric matrix with eigenvalue λ. Prove that the real part of λ is zero, and that $\overline{\lambda}$ is also an eigenvalue.

If $(A - \lambda I)^2 Z = 0$ and $Y = (A - \lambda I)Z$ show, by evaluating $\overline{Y}^t Y$, that $Y = 0$. Hence prove that A satisfies a polynomial equation without repeated roots, and deduce that A is similar to a diagonal matrix.

If x is an eigenvector corresponding to the eigenvalue $\lambda = i\alpha$ and if $u = x + \overline{x}, v = i(x - \overline{x})$ show that

$$Au = \alpha v, \quad Av = -\alpha u.$$

Hence show that A is similar to a diagonal block matrix

$$\begin{bmatrix} 0 & & & & \\ & A_1 & & & \\ & & A_2 & & \\ & & & \ddots & \\ & & & & A_k \end{bmatrix}$$

where each A_i is real and of the form

$$\begin{bmatrix} 0 & \alpha_i \\ -\alpha_i & 0 \end{bmatrix}.$$

1.33 Let V be a vector space of dimension 3 over $\mathbb{R}$ and let $t \in \mathcal{L}(V, V)$ have eigenvalues $-2, 1, 2$. Use the Cayley–Hamilton theorem to express t^{2n} as a real quadratic polynomial in t.

1.34 Let V be a vector space of dimension n over a field F and let $f \in \mathcal{L}(V, V)$ be such that all the zeros of the characteristic polynomial of f lie in F.

Let λ_1 be an eigenvalue of f and let b_1 be an associated eigenvector. Let W be such that $V = Fb_1 \oplus W$ and let $(b'_i)_{2 \leq i \leq n}$ be an ordered basis of W. Show that the matrix of f relative to the basis $\{b_1, b'_2, \ldots, b'_n\}$ is of the form

$$\begin{bmatrix} \lambda_1 & \beta'_{12} & \cdots & \beta'_{1n} \\ & & & \\ 0 & & M & \\ & & & \end{bmatrix}.$$

Observe that in general $\beta'_{12}, \ldots, \beta'_{1n}$ are non-zero, so that W is not f-invariant. Let π be the projection of V onto W and let $g = \pi \circ f$. Show that W is g-invariant and that if g' is the linear transformation induced on W by g then $\mathrm{Mat}\,(g', (b'_i)) = M$. Show also that all the zeros of the characteristic polynomial of g' lie in F.

Deduce that f is *triangularisable* in the sense that there is a basis B of V relative to which the matrix of f is upper triangular with diagonal entries the eigenvalues of f.

1.35 Suppose that $t \in \mathcal{L}(\mathbb{R}^3, \mathbb{R}^3)$ is given by

$$t(a, b, c) = (2a + b - c, -2a - b + 3c, c).$$

Find the eigenvalues and the minimum polynomial of t. Show that t is not diagonalisable. Find a basis of $\mathbb{R}^3$ with respect to which the matrix of t is upper triangular.

1.36 Let $V = \mathbb{Q}_3[X]$ be the vector space of polynomials of degree less than or equal to 2 over the field $\mathbb{Q}$. If $t \in \mathcal{L}(V, V)$ is given by

$$t(1) = -5 - 8X - 5X^2$$
$$t(X) = 1 + X + X^2$$
$$t(X^2) = 4 + 7X + 4X^2,$$

show that t is nilpotent. Find a basis of V with respect to which the matrix of t is upper triangular.

1.37 For each of the following matrices A find a Jordan normal form and an invertible matrix P such that $P^{-1}AP$ is in Jordan normal form.

$$(a) \begin{bmatrix} 39 & -64 \\ 25 & -41 \end{bmatrix}, \quad (b) \begin{bmatrix} -1 & -1 \\ 0 & -1 \end{bmatrix},$$

$$(c) \begin{bmatrix} 1 & 3 & -2 \\ 0 & 7 & -4 \\ 0 & 9 & -5 \end{bmatrix}, \quad (d) \begin{bmatrix} 3 & 0 & 1 \\ 0 & 3 & 0 \\ 0 & 0 & 3 \end{bmatrix}.$$

1.38 Find a Jordan normal form J of the matrix

$$A = \begin{bmatrix} 2 & 1 & 1 & 1 & 0 \\ 0 & 2 & 0 & 0 & 0 \\ 0 & 0 & 2 & 1 & 0 \\ 0 & 0 & 0 & 1 & 1 \\ 0 & -1 & -1 & -1 & 0 \end{bmatrix}.$$

Find also a Jordan basis and hence an invertible matrix P such that $P^{-1}AP = J$.

1.39 For each of the following matrices A find a Jordan normal form J, a Jordan basis, and an invertible matrix P such that $P^{-1}AP = J$.

$$(a) \begin{bmatrix} 22 & -2 & -12 \\ 20 & 0 & -12 \\ 30 & -3 & -16 \end{bmatrix}, \quad (b) \begin{bmatrix} -13 & 8 & 1 & 2 \\ -22 & 13 & 0 & 3 \\ 8 & -5 & 0 & -1 \\ -22 & 13 & 5 & 5 \end{bmatrix}.$$

1.40 Find a Jordan normal form and a Jordan basis for the matrix

$$\begin{bmatrix} 5 & -1 & -3 & 2 & -5 \\ 0 & 2 & 0 & 0 & 0 \\ 1 & 0 & 1 & 1 & -2 \\ 0 & -1 & 0 & 3 & 1 \\ 1 & -1 & -1 & 1 & 1 \end{bmatrix}.$$

1.41 Find the minimum polynomial of the matrix

$$A = \begin{bmatrix} 1 & 0 & -1 & 1 & 0 \\ -4 & 1 & -3 & 2 & 1 \\ -2 & -1 & 0 & 1 & 1 \\ -3 & -1 & -3 & 4 & 1 \\ -8 & -2 & -7 & 5 & 4 \end{bmatrix}.$$

From only the information given by the minimum polynomial, how many essentially different Jordan normal forms are possible? How many linearly independent eigenvectors are there? Does the number of linearly independent eigenvectors determine the Jordan normal form J? If not, does the information given by the minimum polynomial together with the number of linearly independent eigenvectors determine J?

1.42 Find a Jordan normal form of the differentiation map D on the vector space $\mathbb{R}_4[X]$ of polynomials of degree less than or equal to 3 with real coefficients. Find also a Jordan basis for D on $\mathbb{R}_4[X]$.

1.43 If a 3×3 real matrix has eigenvalues $3, 3, 3$ what are the possible Jordan normal forms? Which of these are similar?

1.44 Which of the following are true? If $A, B \in \mathrm{Mat}_{n \times n}(\mathbb{C})$ then AB and BA have the same Jordan normal form

 (i) if A and B are both invertible;
 (ii) if one of A, B is invertible;
 (iii) if and only if A and B are invertible;
 (iv) if and only if one of A, B is invertible.

1.45 Let V be a vector space of dimension n over $\mathbb{C}$. Let $t \in \mathcal{L}(V, V)$ and let λ be an eigenvalue of t. Let J be a matrix that represents t relative to some Jordan basis of V. Show that there are

$$\dim \mathrm{Ker}(t - \lambda \,\mathrm{id}_V)$$

blocks in J with diagonal entries λ.

More generally, if n_i is the number of $i \times i$ blocks with diagonal entries λ and $d_i = \dim \operatorname{Ker}(t - \lambda \operatorname{id}_V)^i$, show that

$$d_i = n_1 + 2n_2 + \cdots + (i - 1)n_{i-1} + i(n_i + n_{i+1} + \ldots).$$

Deduce that $n_i = 2d_i - d_{i-1} - d_{i+1}$.

1.46 Find a Jordan normal form J of the matrix

$$A = \begin{bmatrix} 0 & 1 & 0 & -1 \\ -2 & 3 & 0 & -1 \\ -2 & 1 & 2 & -1 \\ 2 & -1 & 0 & 3 \end{bmatrix}.$$

Find also a Jordan basis and an invertible matrix P such that $P^{-1}AP = J$.

Hence solve the system of differential equations

$$\begin{bmatrix} x_1' \\ x_2' \\ x_3' \\ x_4' \end{bmatrix} = \begin{bmatrix} 0 & 1 & 0 & -1 \\ -2 & 3 & 0 & -1 \\ -2 & 1 & 2 & -1 \\ 2 & -1 & 0 & 3 \end{bmatrix} \begin{bmatrix} x_1 \\ x_2 \\ x_3 \\ x_4 \end{bmatrix}.$$

1.47 Solve each of the following systems of differential equations :

$$(a) \begin{cases} \dfrac{dx_1}{dt} = 5x_1 + 4x_2 \\ \dfrac{dx_2}{dt} = -x_1 \end{cases} \qquad (b) \begin{cases} \dfrac{dx_1}{dt} = 4x_1 - x_2 - x_3 \\ \dfrac{dx_2}{dt} = x_1 + 2x_2 - x_3 \\ \dfrac{dx_3}{dt} = x_1 - x_2 + 2x_2 \end{cases}$$

$$(c) \begin{cases} \dfrac{dx_1}{dt} - 5x_1 + 6x_2 + 6x_3 = 0 \\ \dfrac{dx_2}{dt} + x_1 - 4x_2 - 2x_3 = 0 \\ \dfrac{dx_3}{dt} - 3x_1 + 6x_2 + 4x_3 = 0 \end{cases} \qquad (d) \begin{cases} \dfrac{dx_1}{dt} = x_1 + 3x_2 - 2x_3 \\ \dfrac{dx_2}{dt} = 7x_2 - 4x_3 \\ \dfrac{dx_3}{dt} = 9x_2 - 5x_3 \end{cases}$$

1.48 Solve the system of differential equations

$$\frac{dy_1}{dx} - 3\frac{dy_2}{dx} + 2\frac{dy_3}{dx} = y_1$$

$$-5\frac{dy_2}{dx} + 4\frac{dy_3}{dx} = y_2$$

$$-9\frac{dy_2}{dx} + 7\frac{dy_3}{dx} = y_3$$

1.49 Solve the system of differential equations

$$\frac{dx_1}{dt} = x_1 + x_2$$

$$\frac{dx_2}{dt} = 2x_1 + 3x_2$$

given that $x_1(0) = 0$ and $x_2(0) = 1$.

1.50 Show how the differential equation

$$x''' - 2x'' - 4x' + 8x = 0$$

can be written as a first-order matrix system $X' = AX$. By using the method of the Jordan normal form, solve the equation given the initial conditions

$$x(0) = 0, \quad x'(0) = 0, \quad x''(0) = 16.$$

2: Duality and normal transformations

The *dual* of a vector space V over a field F is the vector space $V^d = \mathcal{L}(V, F)$ of linear functionals $f : V \to F$. If V is of finite dimension and $B = \{v_1, \dots, v_n\}$ is a basis of V then the basis that is *dual* to B is $B^d = \{v_1^d, \dots, v_n^d\}$ where each $v_i^d : V \to F$ is given by

$$v_i^d(v_j) = \begin{cases} 1 & \text{if } i = j; \\ 0 & \text{if } i \neq j. \end{cases}$$

For every $x \in V$ we have

$$x = v_1^d(x)v_1 + v_2^d(x)v_2 + \cdots + v_n^d(x)v_n \,;$$

and for every $f \in V^d$ we have

$$f = f(v_1)v_1^d + f(v_2)v_2^d + \cdots + f(v_n)v_n^d \,.$$

If $(v_i)_n, (w_i)_n$ are ordered bases of V and $(v_i^d)_n, (w_i^d)_n$ the corresponding dual bases then the transition matrix from $(v_i^d)_n$ to $(w_i^d)_n$ is $(P^{-1})^t$ where P is the transition matrix from $(v_i)_n$ to $(w_i)_n$. In particular, consider $V = \mathbb{R}^n$. Note that if

$$B = \{(a_{11}, \dots, a_{1n}), (a_{21}, \dots, a_{2n}), \dots, (a_{n1}, \dots, a_{nn})\}$$

is a basis of $\mathbb{R}^n$ then the transition matrix from B to the canonical basis $(e_i)_n$ of $\mathbb{R}^n$ is $M = [m_{ij}]_{n \times n}$ where $m_{ij} = a_{ji}$. The transition matrix from B^d to $(e_i^d)_n$ is given by $(M^{-1})^t$. We can therefore usefully denote the dual basis by

$$B = \{[\alpha_{11}, \dots, \alpha_{1n}], [\alpha_{21}, \dots, \alpha_{2n}], \dots, [\alpha_{n1}, \dots, \alpha_{nn}]\}$$

where $[\alpha_{i1}, \ldots, \alpha_{in}]$ denotes the ith *row* of M^{-1}, so that

$$[\alpha_{i1}, \ldots, \alpha_{in}](x_1, \ldots, x_n) = \alpha_{i1}x_1 + \cdots + \alpha_{in}x_n.$$

The *bidual* of an element x is $x^\wedge : V^d \to F$ where $x^\wedge(y^d) = y^d(x)$. It is common practice to write $y^d(x)$ as $\langle x, y^d \rangle$ and say that y^d *annihilates* x if $\langle x, y^d \rangle = 0$. For every subspace W of V the set

$$W^\perp = \{y^d \in V^d \mid (\forall x \in W) \, \langle x, y^d \rangle = 0\}$$

is a subspace of W and

$$\dim W + \dim W^\perp = \dim V.$$

The *transpose* of a linear transformation $f : V \to W$ is the linear mapping $f^t : W^d \to V^d$ described by $y^d \mapsto y^d \circ f$. When V is of finite dimension we can identify V and its bidual $(V^d)^d$, in which case we have that $(f^t)^t = f$. Moreover, if $f : V \to W$ is represented relative to fixed ordered bases by the matrix A then $f^t : W^d \to V^d$ is represented relative to the corresponding dual bases by the transpose A^t of A.

If V is a finite-dimensional inner product space then the mapping $\vartheta_V : x \mapsto x^d$ describes a *conjugate isomorphism* from V to V^d, by which we mean that

$$(x + y)^d = x^d + y^d \quad \text{and} \quad (\lambda x)^d = \bar{\lambda} x^d.$$

The *adjoint* $f^\star : W \to V$ of $f : V \to W$ is defined by

$$f^\star = \vartheta_V^{-1} \circ f^t \circ \vartheta_W$$

and is the unique linear transformation such that

$$(\forall x, y \in V) \qquad \langle f(x) | y \rangle = \langle x | f^\star(y) \rangle.$$

We say that f is *normal* if it commutes with its adjoint. If the matrix of f relative to a given ordered basis is A then that of $f^\star$ is $\overline{A}^t$. We say that A is normal if it commutes with $\overline{A}^t$. A matrix is normal if and only if it is unitarily similar to a diagonal matrix, i.e. if there is a matrix U with $U^{-1} = \overline{U}^t$ such that $U^{-1}AU$ is diagonal. A particularly important type of normal transformation occurs when the vector space in question is a real inner product space, and topics dealt with in this section reach as far as the orthogonal reduction of real symmetric matrices and its application to finding the rank and signature of quadratic forms.

2.1   Determine which of the following mappings are linear functionals on the vector space $\mathbb{R}_3[X]$ of all real polynomials of degree less than or equal to 2 :

$$(a)\ f \mapsto f'; \quad (b)\ f \mapsto \int_0^1 f; \quad (c)\ f \mapsto f(2);$$

$$(d)\ f \mapsto f'(2); \quad (e)\ f \mapsto \int_0^1 f^2.$$

2.2   Let $C[0,1]$ be the vector space of continuous functions $f : [0,1] \to \mathbb{R}$. If f_0 is a fixed element of $C[0,1]$, prove that $\varphi : C[0,1] \to \mathbb{R}$ given by

$$\varphi(f) = \int_0^1 f_0(t)\, f(t)\, dt$$

is a linear functional.

2.3   Determine the basis of $(\mathbb{R}^3)^d$ that is dual to the basis

$$\{(1,0,-1),(-1,1,0),(0,1,1)\}$$

of $\mathbb{R}^3$.

2.4   Let $A = \{x_1, x_2\}$ be a basis of a vector space V of dimension 2 and let $A^d = \{\varphi_1, \varphi_2\}$ be the corresponding dual basis of V^d. Find, in terms of φ_1, φ_2 the basis of V^d that is dual to the basis $A' = \{x_1 + 2x_2, 3x_1 + 4x_2\}$ of V.

2.5   Which of the following bases of $(\mathbb{R}^2)^d$ is dual to the basis $\{(-1,2),(0,1)\}$ of $\mathbb{R}^2$?

       $(a)\ \{[-1,2],[0,1]\};$ $(b)\ \{[-1,0],[2,1]\};$
       $(c)\ \{[-1,0],[-2,1]\};$ $(d)\ \{[1,0],[2,-1]\}.$

2.6   (i) Find a basis that is dual to the basis

$$\{(4,5,-2,11),(3,4,-2,6),(2,3,-1,4),(1,1,-1,3)\}$$

of $\mathbb{R}^4$.

     (ii) Find a basis of $\mathbb{R}^4$ whose dual basis is

$$\{[2,-1,1,0],[-1,0,-2,0],[-2,2,1,0],[-8,3,-3,1]\}.$$

2.7   Show that if V is a finite-dimensional vector space over a field F and if A, B are subspaces of V such that $V = A \oplus B$ then $V^d = A^\perp \oplus B^\perp$.
     Is it true that if $V = A \oplus B$ then $V^d = A^d \oplus B^d$?

2: Duality and normal transformations

2.8 Let $\mathbb{R}_3[X]$ be the vector space of polynomials over $\mathbb{R}$ of degree less than or equal to 2. Let t_1, t_2, t_3 be three distinct real numbers and for $i = 1, 2, 3$ define mappings $f_i : \mathbb{R}_3[X] \to \mathbb{R}$ by

$$f_i(p(X)) = p(t_i).$$

Show that $B^d = \{f_1, f_2, f_3\}$ is a basis for the dual space $(\mathbb{R}_3[X])^d$ and determine a basis $B = \{p_1(X), p_2(X), p_3(X)\}$ of $\mathbb{R}_3[X]$ of which B^d is the dual.

2.9 Let $\alpha = (1, 2)$ and $\beta = (5, 6)$ be elements of $\mathbb{R}^2$ and let $\varphi = [3, 4]$ be an element of $(\mathbb{R}^2)^d$. Determine

$(a)\ \alpha^\wedge(\varphi);$ $(b)\ \beta^\wedge(\varphi);$
$(c)\ (2\alpha + 3\beta)^\wedge(\varphi);$ $(d)\ (2\alpha + 3\beta)^\wedge([a, b]).$

2.10 Prove that if S is a subspace of a finite-dimensional vector space V then

$$\dim S + \dim S^\perp = \dim V.$$

If $t \in \mathcal{L}(U, V)$ and $t^d \in \mathcal{L}(V^d, U^d)$ is the dual of t, prove that

$$\mathrm{Ker}\, t^d = (\mathrm{Im}\, t)^\perp.$$

Deduce that if $v \in V$ then one of the following holds :

(i) there exists $u \in U$ such that $t(u) = v$;
(ii) there exists $\varphi \in V^d$ such that $t^d(\varphi) = 0$ and $\varphi(v) = 1$.

Translate these results into a theorem on solving systems of linear equations.

Show that (i) is not satisfied by the system

$$\begin{aligned}
3x + \ y &= 2 \\
x + 2y &= 1 \\
-x + 3y &= 1.
\end{aligned}$$

Find the linear functional φ whose existence is guaranteed by (ii).

2.11 If $s, t : U \to V$ are linear transformations, show that

$$(s \circ t)^d = t^d \circ s^d.$$

Prove that the dual of an injective linear transformation is surjective, and that the dual of a surjective linear transformation is injective.

2.12 Let $t \in \mathcal{L}(\mathbb{R}^3, \mathbb{R}^3)$ be given by the prescription

$$t(a, b, c) = (2a + b, a + b + c, -c).$$

If $X = \{(1,0,0), (1,1,0), (1,1,1)\}$ and $Y^d = \{[1,0,0], [1,1,0], [1,1,1]\}$, find the matrix of t^d with respect to the bases Y^d and X^d.

2.13 Let $\{\alpha_1, \alpha_2, \alpha_3\}$ and $\{\alpha_1, \alpha_2, \alpha_3'\}$ be bases of $\mathbb{R}^3$ that differ only in the third basis element. Suppose that $\{\varphi_1, \varphi_2, \varphi_3\}$ and $\{\varphi_1', \varphi_2', \varphi_3'\}$ are the corresponding dual bases. Prove that φ_3' is a scalar multiple of φ_3.

2.14 Let $C[0, 1]$ denote the space of continuous functions on the interval $[0, 1]$. Given $g \in C[0, 1]$, define $L_g : C[0, 1] \to \mathbb{R}$ by

$$L_g(f) = \int_0^1 f(t) g(t) \, dt.$$

Show that L_g is a linear functional.

Let x be a fixed element of $[0, 1]$ and define $F_x : C[0, 1] \to \mathbb{R}$ by $F_x(f) = f(x)$. Show that F_x is a linear functional. Show also that there is no $g \in C[0, 1]$ such that $F_x = L_g$.

2.15 By a *canonical isomorphism* $\varsigma : V \to V^d$ we mean an isomorphism ς such that, for all $x, y \in V$ and all isomorphisms $f : V \to V$, we have

$$(\star) \qquad\qquad \langle x, \varsigma(y) \rangle = \langle f(x), \varsigma[f(y)] \rangle,$$

where the notation $\langle x, \varsigma(y) \rangle$ means $[\varsigma(y)](x)$.

In this exercise we indicate a proof of the fact that if V is of dimension $n > 1$ over F then there is no canonical isomorphism $\varsigma : V \to V^d$ except when $n = 2$ and F has two elements.

If ς is such an isomorphism show that, for $y \neq 0$, the subspace $\operatorname{Ker} \varsigma(y) = \{\varsigma(y)\}^{\perp}$ is of dimension $n - 1$.

Suppose first that $n > 3$. If there exists $t \in \operatorname{Ker} \varsigma(t)$ for some $t \neq 0$ let $\{t, x_1, \ldots, x_{n-2}\}$ be a basis of $\operatorname{Ker} \varsigma(t)$ and extend this to a basis $\{t, x_1, \ldots, x_{n-2}, z\}$ of V. Let $f : V \to V$ be the (unique) linear transformation such that

$$f(t) = t, \quad f(x_1) = z, \quad f(z) = x_1, \quad \text{and } f(x_i) = x_i \text{ for } i \neq 1.$$

Show that f is an isomorphism that does not satisfy $(\star)$. [*Hint.* Take $x = x_1, y = t$.] If, on the other hand, $t \notin \operatorname{Ker} \varsigma(t)$ for all $t \neq 0$ let

$\{x_1, \ldots, x_{n-1}\}$ be a basis of $\text{Ker}\,\varsigma(t)$ so that $\{x_1, \ldots, x_{n-1}, t\}$ is a basis of V. Show that

$$\{x_1 + x_2, x_2, x_3, \ldots, x_{n-1}, t\}$$

is also a basis of V. Show also that $x_2 \in \text{Ker}\,\varsigma(x_1)$. Now show that if $f : V \to V$ is the (unique) linear transformation such that

$$f(x_1) = x_2, \ f(x_2) = x_1 + x_2, \ f(t) = t, \ f(x_i) = x_i \ (i \neq 1, 2)$$

then f is an isomorphism that does not satisfy $(\star)$. Conclude from these observations that we must have $n = 2$.

Suppose now that F has more than two elements and let $\lambda \in F$ be such that $\lambda \neq 0, 1$. If there exists $t \neq 0$ such that $t \in \text{Ker}\,\varsigma(t)$ observe that $\{t\}$ is a basis of $\text{Ker}\,\varsigma(t)$ and extend this to a basis $\{t, z\}$ of V. If $f : V \to V$ is the (unique) linear transformation such that $f(t) = t, f(z) = \lambda z$ show that f is an isomorphism that does not satisfy $(\star)$. [*Hint.* Take $x = z, y = t$.] If, on the other hand, $t \notin \text{Ker}\,\varsigma(t)$ for all $t \neq 0$ let $\{z\}$ be a basis for $\text{Ker}\,\varsigma(t)$ so that $\{z, t\}$ is a basis for V. If $f : V \to V$ is the (unique) linear transformation such that $f(z) = \lambda z, f(t) = t$ show that f is an isomorphism that does not satisfy $(\star)$. [*Hint.* Take $x = y = z$.] Conclude from these observations that F must have two elements.

Now examine the vector space F^2 where $F = \{0, 1\}$.

[*Hint.* $(F^2)^d$ is the set of linear transformations $f : F \times F \to F$. Since F^2 has four elements there are $2^4 = 16$ laws of composition on F. Only four of these are linear transformations from F^2 to F; and each of these is determined by its action on the natural basis of F^2. Compute $(F^2)^d$ and determine a canonical isomorphism from F^2 onto $(F^2)^d$.]

2.16 Let V be an inner product space of dimension k and let U be a subspace of V of dimension $k - 1$ (a hyperplane). Show that there exists a unit vector n in V such that

$$U = \{x \in V \mid \langle n|x\rangle = 0\}.$$

Given $v \in V$, define

$$v' = v - 2\langle n|v\rangle n.$$

Show that $v - v'$ is orthogonal to U and that $\frac{1}{2}(v + v') \in U$, so that v' is the reflection of v in the hyperplane U. Show also that the mapping $t : V \to V$ defined by

$$t(v) = v'$$

is linear and orthogonal. What can you say about its eigenvalues and eigenvectors?

If $s : \mathbb{R}^3 \to \mathbb{R}^3$ and $t : \mathbb{R}^4 \to \mathbb{R}^4$ are respectively reflections in the plane $3x - y + z = 0$ and in the hyperplane $2x - y + 2z - t = 0$, show that the matrices of s and t are respectively

$$\frac{1}{11} \begin{bmatrix} -7 & 6 & -6 \\ 6 & 9 & 2 \\ -6 & 2 & 9 \end{bmatrix}, \quad \frac{1}{5} \begin{bmatrix} 1 & 2 & -4 & 2 \\ 2 & 4 & 2 & -1 \\ -4 & 2 & 1 & 2 \\ 2 & -1 & 2 & 4 \end{bmatrix}.$$

2.17   Find the equations of the principal axes of the hyperbola

$$-x^2 + 6xy - y^2 = 1.$$

Find also the equations of the principal axes of the ellipsoid

$$7x^2 + 6y^2 + 5z^2 + 4xy - 4yz = 1.$$

2.18   Let V be a finite-dimensional inner product space and let $f : V \to V$ be linear. Show that if A is the matrix of f relative to an orthonormal basis B of V then the matrix of the adjoint f^* of f relative to B is the transpose of the complex conjugate of A.

2.19   For every $A \in \mathrm{Mat}_{n \times n}(\mathbb{C})$ define the *trace* of A by $\mathrm{tr}(A) = \sum_{i=1}^{n} a_{ii}$. Show that if V is the vector space of $n \times n$ matrices over $\mathbb{C}$ then the mapping

$$(A, B) \mapsto \langle A | B \rangle = \mathrm{tr}(B^* A),$$

where B^* denotes the transpose of the complex conjugate of B, is an inner product on V.

Consider, for every $M \in V$, the mapping $f_M : V \to V$ defined by

$$f_M(A) = MA.$$

Show that, relative to the above inner product,

$$(f_M)^* = f_{M^*}.$$

2.20   Let V be a finite-dimensional inner product space. Show that for every $f \in V^d$ there is a unique $\beta \in V$ such that

$$(\forall x \in V) \qquad f(x) = \langle x | \beta \rangle.$$

2: Duality and normal transformations

[*Hint*. Let $\{\alpha_1, \ldots, \alpha_n\}$ be an orthonormal basis of V and consider

$$\beta = \sum_{i=1}^{n} \overline{f(\alpha_i)}\, \alpha_i. \,]$$

Show as follows that this result does not necessarily hold for inner product spaces of infinite dimension. Let V be the vector space of polynomials over $\mathbb{C}$. Show that the mapping

$$(p, q) \mapsto \langle p|q \rangle = \int_0^1 p(t)\, \overline{q(t)}\, dt$$

is an inner product on V. Let z be a fixed element of $\mathbb{C}$ and let $f \in V^d$ be the 'evaluation at z' map given by

$$(\forall p \in V) \qquad f(p) = p(z).$$

Show that there is no $q \in V$ such that $(\forall p \in V)\, f(p) = \langle p|q \rangle$.
[*Hint*. Suppose that such a q exists. Let $r \in V$ be given by $r(t) = t - z$ and show that, for every $p \in V$,

$$0 = \int_0^1 r(t)\, p(t)\, \overline{q(t)}\, dt.$$

Now let p be given by $p(t) = \overline{r(t)}q(t)$ and deduce the contradiction $q = 0$.]

For the rest of this question let V continue to be the vector space of polynomials over $\mathbb{C}$ with the above inner product. If $p \in V$ is given by $p(t) = \sum a_k t^k$ define $\overline{p} \in V$ by $\overline{p}(t) = \sum \overline{a}_k t^k$, and let $f_p : V \to V$ be given by

$$(\forall q \in V) \qquad f_p(q) = pq$$

where, as usual, $(pq)(t) = p(t)q(t)$. Show that $(f_p)^\star$ exists and is $f_{\overline{p}}$.

Now let $D : V \to V$ be the differentiation map. Show that D does not admit an adjoint.
[*Hint*. Suppose that $D^\star$ exists and show that, for all $p, q \in V$,

$$\langle p \,|\, D(q) + D^\star(q) \rangle = p(1)\overline{q}(1) - p(0)\overline{q}(0).$$

Suppose now that q is a fixed element of V such that $q(0) = 0$ and $q(1) = 1$. Use the previous part of the question (with $z = 1$) to obtain the required contradiction.]

2.21 Let $C[0,1]$ be the inner product space of real continuous functions on $[0,1]$ with the integral inner product. Let $K : C[0,1] \to C[0,1]$ be the integral operator defined by

$$K(f) = \int_0^1 xy\, f(y)\, dy.$$

Prove that K is self-adjoint.

For every positive integer n let f_n be given by

$$f_n(x) = x^n - \frac{2}{n+2}.$$

Show that f_n is an eigenfunction of K with associated eigenvalue 0. Use the Gram–Schmidt orthonormalisation process to find two orthogonal eigenfunctions of K with associated eigenvalue 0.

Prove that K has only one non-zero eigenvalue. Find this eigenvalue and an associated eigenfunction.

2.22 Let t be a skew-adjoint transformation on a unitary space V. Prove that $\mathrm{id} \pm t$ is a bijection and that the transformation

$$s = (\mathrm{id} - t)(\mathrm{id} + t)^{-1}$$

is unitary. Show also that s cannot have -1 as an eigenvalue.

2.23 If S is a real symmetric matrix and T is a real skew-symmetric matrix of the same order, show that

$$\det(I - T - iS) \neq 0.$$

Show also that the matrix

$$U = (I + T + iS)(I - T - iS)^{-1}$$

is unitary.

2.24 Let A be a real symmetric matrix and let S be a real skew-symmetric matrix of the same order. Suppose that A and S commute and that $\det(A - S) \neq 0$. Prove that

$$(A + S)(A - S)^{-1}$$

is orthogonal.

2: Duality and normal transformations

2.25 A complex matrix A is such that $\overline{A}^t A = -A$. Show that the eigenvalues of A are either 0 or -1.

2.26 Let A and B be orthogonal $n \times n$ matrices with $\det A = -\det B$. Prove that $A + B$ is singular.

2.27 Let A be an orthogonal $n \times n$ matrix. Prove that

(1) if $\det A = 1$ and n is odd, or if $\det A = -1$ and n is even, then 1 is an eigenvalue of A;

(2) if $\det A = -1$ then -1 is an eigenvalue of A.

2.28 If A is a skew-symmetric matrix and $g(X)$ is a polynomial such that $g(A) = 0$, prove that $g(-A) = 0$. Deduce that the minimum polynomial of A contains only terms of even degree.

Deduce that if A is skew-symmetric and $f(X), g(X)$ are polynomials whose terms are respectively odd and even then $f(A), g(A)$ are respectively skew-symmetric and symmetric.

2.29 For every complex $n \times n$ matrix A let

$$N(A) = \operatorname{tr}(\overline{A}^t A) = \sum_{i=1}^{n} [\overline{A}^t A]_{ii}.$$

Prove that, for every unitary $n \times n$ matrix U,

$$N(UA) = N(AU) = N(A) \quad \text{and} \quad N(A - U) = N(I_n - U^{-1}A).$$

2.30 If the matrix A is normal and non-singular prove that so is A^{-1}.

Prove that $\overline{A}^t = p(A)$ for some polynomial $p(X)$ if and only if A is normal.

2.31 Prove that if A is a normal matrix and $g(X)$ is any polynomial then $g(A)$ is normal.

2.32 If A and B are real symmetric matrices prove that $A + iB$ is normal if and only if A, B commute.

2.33 Let A be a real skew-symmetric $n \times n$ matrix. Show that $\det(-A) = (-1)^n \det A$ and deduce thst if n is odd then $\det A = 0$. Show also that every quadratic form $x^t A x$ is identically zero.

Prove that the non-zero eigenvalues of A are of the form $i\mu$ where $\mu \in \mathbb{R}$. If $x = y + iz$ where $y, z \in \mathbb{R}^n$ is an eigenvector associated with the eigenvalue $i\mu$, show that $Ay = -\mu z$ and $Az = \mu y$. Show also that $y^t y = z^t z$ and that $y^t z = 0$. If $Au = 0$ show also that $u^t y = u^t z = 0$.

Find the eigenvalues of the matrix

$$A = \begin{bmatrix} 0 & 2 & -2 \\ -2 & 0 & -1 \\ 2 & 1 & 0 \end{bmatrix}$$

and an orthogonal matrix P such that

$$P^t A P = \begin{bmatrix} 0 & 0 & 0 \\ 0 & 0 & 3 \\ 0 & -3 & 0 \end{bmatrix}.$$

2.34 Consider the quadratic form $q(x) = x^t A x$ on $\mathbb{R}^n$. Prove that $q(x) \geq 0$ for all $x \in \mathbb{R}^n$ if and only if the rank of q equals the signature of q. Prove also that $q(x) \geq 0$ for all $x \in \mathbb{R}^n$ with $q(x) = 0$ only when $x = 0$ if and only if the rank and signature of q are each n.

2.35 With respect to the standard basis for $\mathbb{R}^3$, a quadratic form q is represented by the matrix

$$A = \begin{bmatrix} 1 & 1 & -1 \\ 1 & 1 & 0 \\ -1 & 0 & -1 \end{bmatrix}.$$

Is q positive definite? Is q positive semi-definite? Find a basis of $\mathbb{R}^3$ with respect to which the matrix representing q is in normal form.

2.36 Let f be the bilinear form on $\mathbb{R}^2 \times \mathbb{R}^2$ given by

$$f((x_1, x_2), (y_1, y_2)) = x_1 y_1 + x_1 y_2 + 2x_2 y_1 + x_2 y_2.$$

Find a symmetric bilinear form g and a skew-symmetric bilinear form h such that $f = g + h$.

Let q be the quadratic form given by $q(x) = f(x, x)$ where $x \in \mathbb{R}^2$. Find the matrix of q with respect to the standard basis. Find also the rank and signature of q. Is q positive definite? Is q positive semi-definite?

2.37 Write the quadratic form

$$4x^2 + 4y^2 + 4z^2 - 2yz + 2xz - 2xy$$

in matrix notation and show that there is an orthogonal transformation $(x, y, z) \mapsto (u, v, w)$ which transforms the quadratic form to

$$3u^2 + 3v^2 + 6w^2.$$

Deduce that the original form is positive definite.

2: Duality and normal transformations

2.38 By completing squares, find the rank and signature of the following quadratic forms :

(1) $2y^2 - z^2 + xy + xz$;

(2) $2xy - xz - yz$;

(3) $yz + xz + xy + xt + yt + zt$.

2.39 For each of the following quadratic forms write down the symmetric matrix A for which the form is expressible as $x^t Ax$. Diagonalise each of the forms and in each case find a real non-singular matrix P for which the matrix $P^t AP$ is diagonal with entries in $\{1, -1, 0\}$.

(1) $x^2 + 2y^2 + 9z^2 - 2xy + 4xz - 6yz$;

(2) $4xy + 2yz$;

(3) $x^2 + 4y^2 + z^2 - 4t^2 + 2xy - 2xt + 6yz - 8yt - 14zt$.

2.40 Find the rank and signature of the quadratic form

$$Q(x_1, \ldots, x_n) = \sum_{r<s} (x_r - x_s)^2.$$

2.41 Show that the rank and signature of the quadratic form

$$\sum_{r,s=1}^{n} (\lambda rs + r + s) x_r x_s$$

are independent of λ.

2.42 Let A be the matrix associated with the quadratic form $Q(x_1, \ldots, x_n)$ and let λ be an eigenvalue of A. Show that there exist $a_1, \ldots, a_n$ not all zero such that

$$Q(a_1, \ldots, a_n) = \lambda(a_1^2 + \cdots + a_n^2).$$

2.43 If the real square matrix A is such that $\det A \neq 0$ show that the quadratic form $x^t A^t Ax$ is positive definite.

2.44 Let $f : \mathbb{R}^n \times \mathbb{R}^n \to \mathbb{R}$ be a symmetric bilinear form and let Q_f be the associated quadratic form. Suppose that Q_f is positive definite and let $g : \mathbb{R}^n \times \mathbb{R}^n \to \mathbb{R}$ be a symmetric bilinear form with associated quadratic form Q_g. Prove that there is a basis of $\mathbb{R}^n$ with respect to which Q_f and Q_g are each represented by sums of squares.

For every $x \in \mathbb{R}^n$ let $f_x \in (\mathbb{R}^n)^d$ be given by $f_x(y) = f(x, y)$. Call f *degenerate* if there exists $x \in \mathbb{R}^n$ with $f_x = 0$. Determine the scalars $\lambda \in \mathbb{R}$ such that $g - \lambda f$ is degenerate. Show that such scalars are the roots of the equation $\det(B - \lambda A) = 0$ where A, B represent f, g relative to some basis of $\mathbb{R}^n$.

By considering the quadratic forms $2xy + 2yz$ and $x^2 - y^2 + 2xz$ show that the result in the first paragraph fails if neither f nor g is positive definite.

2.45 Evaluate

$$\int_{-\infty}^{\infty} \int_{-\infty}^{\infty} \int_{-\infty}^{\infty} e^{-(x^2 + y^2 + z^2 + xy + xz + yz)} \, dx \, dy \, dz \, .$$

Solutions to Chapter 1

1.1

(i) False. For example, take $b = -a_1$.

(ii) True.

(iii) False. $\{(1,1,1)\}$ is a basis, so the dimension is 1.

(iv) False. For example, take $A = \{0\}$ or $A = \{v, 2v\}$.

(v) True.

(vi) False. For example, take $A = \mathbb{R}^n$.

(vii) True.

(viii) False. $\{(x, \lambda x) \mid x \in \mathbb{R}\}$ is a subspace of $\mathbb{R}^2$ for every $\lambda \in \mathbb{R}$.

(ix) True.

(x) True.

(xi) False. An isomorphism is always represented by a *non-singular* matrix.

(xii) False. Consider, for example, $\mathbb{R}^2$ and $\mathbb{C}^2$. The statement is true, however, if the vector spaces have the same ground field.

(xiii) False. $\begin{bmatrix} 0 & 1 \\ 1 & 0 \end{bmatrix}$ is a counter-example.

(xiv) False. For example,

$$\begin{bmatrix} 1 & 0 \\ 0 & 0 \end{bmatrix}\begin{bmatrix} 1 & 0 \\ 1 & 0 \end{bmatrix} = \begin{bmatrix} 1 & 3 \\ 0 & 4 \end{bmatrix}\begin{bmatrix} 1 & 0 \\ 0 & 0 \end{bmatrix}.$$

(xv) True.

(xvi) True.

(xvii) False. Take, for example, $f, g : \mathbb{R}^n \to \mathbb{R}^n$ given by $f(x, y) = (0, 0)$ and $g(x, y) = (x, y)$. Relative to the standard basis of $\mathbb{R}^n$ we see that f is represented by the zero matrix and g is represented by the identity matrix; and there is no invertible matrix P such that $P^{-1}I_2P = 0$.

(xviii) True.

(xix) False. The transformation t is non-singular (an isomorphism), but $\begin{bmatrix} 1 & 2 \\ 1 & 2 \end{bmatrix}$ is singular.

(xx) False. The matrix $\begin{bmatrix} 1 & 1 \\ 0 & 1 \end{bmatrix}$ is not diagonalisable.

1.2   We have that

$$(a,b,c) \in \operatorname{Ker} t_1 \iff (a+b, b+c, c+a) = (0,0,0)$$
$$\iff a = b = c = 0$$

and so $\operatorname{Ker} t_1 = \{0\}$. It follows from the dimension theorem that $\operatorname{Im} t_1 = \mathbb{R}^3$.

As for t_2, we have

$$(a,b,c) \in \operatorname{Ker} t_2 \iff a - b = 0, b - c = 0$$
$$\iff a = b = c$$

and so $\operatorname{Ker} t_2 = \{(a,a,a) \mid a \in \mathbb{R}\}$. It is clear from the definition of t_2 that $\operatorname{Im} t_2 = \{(a,b,0) \mid a,b \in \mathbb{R}\}$.

Likewise, it is readily seen that
$\operatorname{Ker} t_3 = \{0\}, \quad \operatorname{Im} t_3 = \mathbb{R}^3,$
$\operatorname{Ker} t_4 = \{(0,0,a) \mid a \in \mathbb{R}\}, \quad \operatorname{Im} t_4 = \{(a,b,b) \mid a,b \in \mathbb{R}\}.$
If $\operatorname{Ker} t_i \cap \operatorname{Im} t_i = \{0\}$ then by the dimension theorem we have

$$\dim(\operatorname{Ker} t_i + \operatorname{Im} t_i) = \dim \mathbb{R}^3$$

and so $\operatorname{Ker} t_i + \operatorname{Im} t_i = \mathbb{R}^3$. Now for $i = 1, 2, 3, 4$ we have from the above that $\operatorname{Ker} t_i \cap \operatorname{Im} t_i = \{0\}$. Thus we see that $\mathbb{R}^3 = \operatorname{Ker} t_i \oplus \operatorname{Im} t_i$ holds in all cases.

$\operatorname{Im} t_2$ is t_3–invariant. For, if $v \in \operatorname{Im} t_2$ then $v = (a,b,0)$ and so

$$t_3(v) = t_3(a,b,0) = (-b,a,0) \in \operatorname{Im} t_2.$$

However, $\operatorname{Ker} t_2$ is not t_3–invariant. For $(1,1,1) \in \operatorname{Ker} t_2$ but $t_3(1,1,1) = (-1,1,1) \notin \operatorname{Ker} t_2$.

For the last part, we have that $(t_3 \circ t_4)(a, b, c) = (-b, a, b)$ and that $(t_4 \circ t_3)(a, b, c) = (-b, a, a)$. Consequently,

$$\mathrm{Ker}(t_3 \circ t_4) = \{(0, 0, a) \mid a \in \mathbb{R}\};$$
$$\mathrm{Im}(t_3 \circ t_4) = \{(-b, a, b) \mid a, b \in \mathbb{R}\};$$
$$\mathrm{Ker}(t_4 \circ t_3) = \{(0, 0, a) \mid a \in \mathbb{R}\};$$
$$\mathrm{Im}(t_4 \circ t_3) = \{(-b, a, a) \mid a, b \in \mathbb{R}\}.$$

1.3 Reducing the matrix to row-echelon form we obtain

$$
\begin{bmatrix} 3 & -1 & 1 \\ -1 & 5 & -1 \\ 1 & -1 & 3 \end{bmatrix}
\longrightarrow
\begin{bmatrix} 1 & -1 & 3 \\ -1 & 5 & -1 \\ 3 & -1 & 1 \end{bmatrix}
\longrightarrow
\begin{bmatrix} 1 & -1 & 3 \\ 0 & 4 & 2 \\ 0 & 2 & -8 \end{bmatrix}
$$

$$
\longrightarrow
\begin{bmatrix} 1 & -1 & 3 \\ 0 & 2 & -8 \\ 0 & 4 & 2 \end{bmatrix}
\longrightarrow
\begin{bmatrix} 1 & -1 & 3 \\ 0 & 2 & -8 \\ 0 & 0 & 18 \end{bmatrix}.
$$

Note that we have been careful not to divide by any number that is divisible by either 2 or 3 (since these will be zero in $\mathbb{Z}_2$ and $\mathbb{Z}_3$ respectively).

(i) When $F = \mathbb{R}$ the rank of the row echelon matrix is 3, in which case $\dim \mathrm{Im}\, t = 3$ and hence $\dim \mathrm{Ker}\, t = 0$.

(ii) When $F = \mathbb{Z}_2$ we have that $2, 18, -8$ are zero so that the rank is 1, in which case $\dim \mathrm{Im}\, t = 1$ and $\dim \mathrm{Ker}\, t = 2$.

(iii) When $F = \mathbb{Z}_3$ we have that 18 is zero so that the rank is 2, in which case $\dim \mathrm{Im}\, t = 2$ and $\dim \mathrm{Ker}\, t = 1$.

$V = \mathrm{Ker}\, t \oplus \mathrm{Im}\, t$ holds in cases (i) and (ii), but not in case (iii); for in case (iii) we have that $(1, 1, 1)$ belongs to both $\mathrm{Ker}\, t$ and $\mathrm{Im}\, t$.

1.4 If $s \circ t = \mathrm{id}_V$ then s is surjective, hence bijective (since V is of finite dimension). Then $t = s^{-1}$ and so $t \circ s = \mathrm{id}_V$.

Suppose that W is t–invariant, so that $t(W) \subseteq W$. Since t is an isomorphism we must have $\dim t(W) = \dim W$ and so $t(W) = W$. Hence $W = s[t(W)] = s(W)$ and W is s–invariant.

The result is false for infinite-dimensional spaces. For example, consider the real vector space $\mathbb{R}[X]$ of polynomials over $\mathbb{R}$. Let s be the differentiation map and t the integration map. We have $s \circ t = \mathrm{id}$ but $t \circ s \neq \mathrm{id}$.

1.5 $\mathrm{Ker}\, D = \{a \mid a \in F\}$ and $\mathrm{Im}\, D = \{p(X) \mid \deg p(X) \le n - 2\}$. Clearly, $\mathrm{Im}\, D$ is isomorphic to V_{n-1} and $\mathrm{Ker}\, D$ is isomorphic to F. Now $\mathrm{Ker}\, D \cap \mathrm{Im}\, D \ne \{0\}$ since if $a \in F$ with $a \ne 0$ then the constant polynomial a belongs to both.

The same results do not hold when the ground field is $\mathbb{Z}_2$. For example, in this case we see that the polynomial X^2 belongs to the kernel of D.

1.6   Let $s, t \in \mathcal{L}(V, V)$. Then if $w \in \mathrm{Im}(s \circ t)$ we have $w = s[t(u)]$ for some $u \in V$ which shows that $w \in \mathrm{Im}\, s$. Thus $\mathrm{Im}(s \circ t) \subseteq \mathrm{Im}\, s$. The first chain now follows by taking $s = t^n$.

Similarly, if $u \in \mathrm{Ker}\, t^n$ then $s[t^n(u)] = s(0) = 0$ gives $u \in \mathrm{Ker}(s \circ t^n)$ and so $\mathrm{Ker}\, t^n \subseteq \mathrm{Ker}(s \circ t^n)$. The second chain now follows by taking $s = t$.

Now we cannot have an infinite number of strict inclusions in the first chain since $X \subset Y$ implies that $\dim X < \dim Y$, and the dimension of V is finite. Hence the chain is finite. It follows that there exists a positive integer p such that $\mathrm{Im}\, t^p = \mathrm{Im}\, t^{p+k}$ for all positive integers k. Since $\dim \mathrm{Im}\, t^p + \dim \mathrm{Ker}\, t^p = \dim V$ the corresponding results for the kernel chain are easily deduced.

To show that $V = \mathrm{Im}\, t^p \oplus \mathrm{Ker}\, t^p$ it suffices, by the dimension argument, to prove that $\mathrm{Im}\, t^p \cap \mathrm{Ker}\, t^p = \{0\}$. Now if $x \in \mathrm{Im}\, t^p \cap \mathrm{Ker}\, t^p$ then $t^p(x) = 0$ and there exists $v \in V$ such that $x = t^p(v)$. Consequently

$$0 = t^p(x) = t^{2p}(v)$$

and so $v \in \mathrm{Ker}\, t^{2p} = \mathrm{Ker}\, t^p$ whence $x = t^p(v) = 0$.

For the last part, observe that if $x \in \mathrm{Im}\, t^p$ then $x = t^p(v)$ gives $t(x) = t^{p+1}(v) \in \mathrm{Im}\, t^{p+1} \subseteq \mathrm{Im}\, t^p$ and so $\mathrm{Im}\, t^p$ is t–invariant. Also, if $x \in \mathrm{Ker}\, t^p$ then $t^p(x) = 0$ gives $t^{p+1}(x) = 0$ so $t^p[t(x)] = 0$ whence $t(x) \in \mathrm{Ker}\, t^p$ and so $\mathrm{Ker}\, t^p$ is t–invariant.

1.7   If $f \circ f = 0$ then $(\forall x \in V)\ f(x) \in \mathrm{Ker}\, f$ and so $\mathrm{Im}\, f \subseteq \mathrm{Ker}\, f$.
We know that

$$n = \dim V = \dim \mathrm{Im}\, f + \dim \mathrm{Ker}\, f = r + \dim \mathrm{Ker}\, f$$

and, by the above, $\dim \mathrm{Ker}\, f \ge \dim \mathrm{Im}\, f = r$. Hence $2r \le n$.

If W is a subspace such that $V = \mathrm{Ker}\, f \oplus W$ then we have that $\dim V = \dim \mathrm{Ker}\, f + \dim W$ and so

$$\dim W = \dim V - \dim \mathrm{Ker}\, f = \dim \mathrm{Im}\, f = r.$$

If $\{w_1, \ldots, w_r\}$ is a basis of W then for $i = 1, \ldots, r$ we have $f(w_i) \in$ Im $f \subseteq$ Ker f. Moreover, $\{f(w_1), \ldots, f(w_r)\}$ is linearly independent since

$$\sum_{i=1}^{r} \lambda_i f(w_i) = 0 \Rightarrow f\left(\sum_{i=1}^{r} \lambda_i w_i\right) = 0$$

$$\Rightarrow \sum_{i=1}^{r} \lambda_i w_i \in \text{Ker } f$$

$$\Rightarrow \sum_{i=1}^{r} \lambda_i w_i \in \text{Ker } f \cap W = \{0\}$$

$$\Rightarrow \sum_{i=1}^{r} \lambda_i w_i = 0$$

$$\Rightarrow (i = 1, \ldots, r) \; \lambda_i = 0.$$

Every linearly independent subset of a vector space can be enlarged to form a basis so, since dim Ker $f = n - r$, we can enlarge the independent subset $\{f(w_1), \ldots, f(w_r)\}$ of Ker f to form a basis of Ker f. Thus we may choose $n - 2r$ elements $x_1, \ldots, x_{n-2r}$ of Ker f such that

$$\{f(w_1), \ldots, f(w_r), x_1, \ldots, x_{n-2r}\}$$

is a basis for Ker f. Since $V = W \oplus \text{Ker } f$ it follows that

$$\{w_1, \ldots, w_r, f(w_1), \ldots, f(w_r), x_1, \ldots, x_{n-2r}\}$$

is a basis for V.

Using the fact that $f \circ f = 0$ and each $x_i \in \text{Ker } f$ it is readily seen that the matrix of f relative to this basis is of the form

$$\begin{bmatrix} 0_r & 0 \\ I_r & 0 \\ 0 & 0 \end{bmatrix}.$$

Suppose now that A is a non-zero $n \times n$ matrix over F. If $A^2 = 0$ and if $f : V \to V$ is represented by A relative to some fixed ordered basis then $f \circ f = 0$ and, from the above, there is a basis of V with respect to which the matrix of f is of the above form. Thus A is similar to this matrix. Conversely, if M denotes the above matrix then clearly $M^2 = 0$. So if A is similar to M there is an invertible matrix P such that $A = P^{-1}MP$ whence $A^2 = P^{-1}M^2P = P^{-1}0P = 0$.

1.8   (1) Sums and scalar multiples of elements of V_1, V_2 are clearly elements of V_1, V_2 respectively.

(2) If $x \in V_1$ then $x = x_1(b_1 + b_4) + x_2(b_2 + b_3)$ shows that V_1 is generated by $\{b_1 + b_4, b_2 + b_3\}$. Also, if $x_1(b_1 + b_4) + x_2(b_2 + b_3) = 0$ then, since $\{b_1, b_2, b_3, b_4\}$ is a basis of V, we have $x_1 = x_2 = 0$. Thus $\{b_1 + b_4, b_2 + b_3\}$ is a basis of V_1. Similarly, $\{b_1 - b_4, b_2 - b_3\}$ is a basis of V_2.

(3) It is clear from the definitions of V_1 and V_2 that we have $V_1 \cap V_2 = \{0\}$. Consequently, the sum $V_1 + V_2$ is direct. Since V_1, V_2 are of dimension 2 and V is of dimension 4, it follows that $V = V_1 \oplus V_2$.

(4) To find the matrix of id_V relative to the bases $B = \{b_1, b_2, b_3, b_4\}$ and $C = \{b_1 + b_4, b_2 + b_3, b_2 - b_3, b_1 - b_4\}$ we observe that

$$b_1 = \tfrac{1}{2}(b_1 + b_4) + 0(b_2 + b_3) + 0(b_2 - b_3) + \tfrac{1}{2}(b_1 - b_4)$$
$$b_2 = 0(b_1 + b_4) + \tfrac{1}{2}(b_2 + b_3) + \tfrac{1}{2}(b_2 - b_3) + 0(b_1 - b_4)$$
$$b_3 = 0(b_1 + b_4) + \tfrac{1}{2}(b_2 + b_3) - \tfrac{1}{2}(b_2 - b_3) + 0(b_1 - b_4)$$
$$b_4 = \tfrac{1}{2}(b_1 + b_4) + 0(b_2 + b_3) + 0(b_2 - b_3) - \tfrac{1}{2}(b_1 - b_4).$$

The matrix in question is therefore

$$A = \begin{bmatrix} \tfrac{1}{2} & 0 & 0 & \tfrac{1}{2} \\ 0 & \tfrac{1}{2} & \tfrac{1}{2} & 0 \\ 0 & \tfrac{1}{2} & -\tfrac{1}{2} & 0 \\ \tfrac{1}{2} & 0 & 0 & -\tfrac{1}{2} \end{bmatrix}.$$

It is readily seen that

$$A^{-1} = 2A = \begin{bmatrix} 1 & 0 & 0 & 1 \\ 0 & 1 & 1 & 0 \\ 0 & 1 & -1 & 0 \\ 1 & 0 & 0 & -1 \end{bmatrix}.$$

Suppose now that M is centro-symmetric; i.e. of the form

$$\begin{bmatrix} a & b & c & d \\ e & f & g & h \\ h & g & f & e \\ d & c & b & a \end{bmatrix}.$$

36

Let f represent M relative to the basis B. Then the matrix of f relative to the basis C is given by AMA^{-1}, which is readily seen to be of the form

$$K = \begin{bmatrix} \alpha & \beta & 0 & 0 \\ \gamma & \delta & 0 & 0 \\ 0 & 0 & \epsilon & \varsigma \\ 0 & 0 & \eta & \vartheta \end{bmatrix}.$$

Thus if M is centro-symmetric it is similar to a matrix of the form K.

1.9 If F is not of characteristic 2 then $1_F + 1_F \neq 0_F$. Writing $2 = 1_F + 1_F$ we have that $\frac{1}{2} \in F$. Given $x \in V$ we then observe that

$$
\begin{aligned}
x &= \tfrac{1}{2}x + \tfrac{1}{2}f(x) + \tfrac{1}{2}x - \tfrac{1}{2}f(x) \\
&= \tfrac{1}{2}x + f(\tfrac{1}{2}x) + \tfrac{1}{2}x - f(\tfrac{1}{2}x) \\
&= (\mathrm{id}_V + f)(\tfrac{1}{2}x) + (\mathrm{id}_V - f)(\tfrac{1}{2}x)
\end{aligned}
$$

so $V = \mathrm{Im}(\mathrm{id}_V + f) + \mathrm{Im}(\mathrm{id}_V - f)$. Also, if $x \in \mathrm{Im}(\mathrm{id}_V + f) \cap \mathrm{Im}(\mathrm{id}_V - f)$ then $x = y + f(y) = z - f(z)$ for some $y, z \in V$ and hence, since $f \circ f = \mathrm{id}_V$ by hypothesis,

$$
\begin{aligned}
f(x) &= f(y) + f[f(y)] = f(y) + y = x; \\
f(x) &= f(z) - f[f(z)] = f(z) - z = -x,
\end{aligned}
$$

whence $x = 0$. Thus $V = \mathrm{Im}(\mathrm{id}_V + f) \oplus \mathrm{Im}(\mathrm{id}_V - f)$.

If $A^2 = I_n$, let f represent A relative to some fixed ordered basis. Then $f \circ f = \mathrm{id}_V$. Let $\{a_1, \ldots, a_p\}$ be a basis of $\mathrm{Im}(\mathrm{id}_V + f)$ and $\{a_{p+1}, \ldots, a_n\}$ be a basis of $\mathrm{Im}(\mathrm{id}_V - f)$. Then $\{a_1, \ldots, a_n\}$ is a basis of V. Now since $a_1 = b + f(b)$ for some $b \in V$ we have $f(a_1) = f(b) + f[f(b)] = f(b) + b = a_1$, and similarly for $a_2, \ldots, a_p$. Likewise, $a_{p+1} = c - f(c)$ for some $c \in V$ so $f(a_{p+1}) = f(c) - f[f(c)] = f(c) - c = -a_{p+1}$, and similarly for $a_{p+2}, \ldots, a_n$. Hence the matrix of f relative to the basis $\{a_1, \ldots, a_n\}$ is

$$\Delta_p = \begin{bmatrix} I_p & 0 \\ 0 & -I_{n-p} \end{bmatrix},$$

and A is then similar to this matrix. Conversely, if A is similar to a matrix of the form Δ_p then there is an invertible matrix Q such that $Q^{-1}AQ = \Delta_p$. Then

$$A^2 = (Q\Delta_p Q^{-1})^2 = Q\Delta_p^2 Q^{-1} = QI_n Q^{-1} = I_n.$$

Suppose now that F is of characteristic 2, so that $x + x = 0$ and hence $x = -x$ for every $x \in F$. Let $f \circ f = \mathrm{id}_V$ and let $g = \mathrm{id}_V + f$. Then

$$(\star) \qquad g(x) = 0 \iff x + f(x) = 0 \iff x = -f(x) = f(x).$$

Moreover, for every $x \in V$ we have $g[g(x)] = g[x + f(x)] = x + f(x) + f[x + f(x)] = x + f(x) + f(x) + f[f(x)] = x + f(x) + f(x) + x = 0$ and hence $g \circ g = 0$.

Suppose now that $A^2 = I_n$ and let f represent A relative to some fixed ordered basis. Let $g = \mathrm{id}_V + f$ and note from the above that $\mathrm{Im}\, g \subseteq \mathrm{Ker}\, g$. Let $\{g(c_1), \ldots, g(c_p)\}$ be a basis of $\mathrm{Im}\, g$ and extend this to a basis

$$\{b_1, \ldots, b_{n-2p}, g(c_1), \ldots, g(c_p)\}$$

of $\mathrm{Ker}\, g$ (which is of dimension $n - \dim \mathrm{Im}\, g = n - p$). Consider now the set

$$B = \{b_1, \ldots, b_{n-2p}, g(c_1), c_1, \ldots, g(c_p), c_p\}.$$

This set has n elements; for $c_i = b_j$ gives the contradiction $g(c_i) = g(b_j) = 0$, and $c_i = g(c_j)$ gives the contradiction $g(c_i) = g[g(c_j)] = 0$. It is also linearly independent; for if we had

$$\sum \lambda_i b_i + \sum \mu_j g(c_j) + \sum \nu_j c_j = 0$$

then, applying g and using the fact that $b_i, g(c_j) \in \mathrm{Ker}\, g$, we deduce that $\sum \nu_j g(c_j) = 0$ whence each $\nu_j = 0$, and then from $\sum \lambda_i b_i + \sum \mu_j g(c_j) = 0$ we obtain $\lambda_i = 0 = \mu_j$ for all i, j. Thus B is a basis of V. To compute the matrix of f relative to the basis B we observe that since $b_i \in \mathrm{Ker}\, g$ we have, by $(\star)$, that $f(b_i) = b_i$ for every i. Also, $f[g(c_i)] = g[g(c_i)] + g(c_i) = g(c_i)$ so that we have

$$f[g(c_i)] = 1.g(c_i) + 0.c_i \; ;$$
$$f(c_i) = 1.g(c_i) + 1.c_i \; .$$

It follows from these observations that the matrix of f relative to B is

$$\nabla_p = \begin{bmatrix} I_{n-2p} & & & & & & & \\ & 1 & 1 & & & & & \\ & 0 & 1 & & & & & \\ & & & 1 & 1 & & & \\ & & & 0 & 1 & & & \\ & & & & & \ddots & & \\ & & & & & & 1 & 1 \\ & & & & & & 0 & 1 \end{bmatrix}.$$

Consequently A is similar to this matrix. Conversely, if A is similar to a matrix of the form ∇_p then there is an invertible matrix Q such that $Q^{-1}AQ = \nabla_p$ and so

$$A^2 = (Q\nabla_p Q^{-1})^2 = Q\nabla_p^2 Q^{-1} = QI_n Q^{-1} = I_n.$$

1.10 If $t \in \mathcal{L}(\mathbb{R}^2, \mathbb{R}^2)$ is given by $t(a, b) = (b, 0)$ then clearly $\operatorname{Im} t = \operatorname{Ker} t \neq \{0\}$.

If $t \in \mathcal{L}(\mathbb{R}^3, \mathbb{R}^3)$ is given by $t(a, b, c) = (c, 0, 0)$ then $\operatorname{Im} t \subset \operatorname{Ker} t$.

If $t \in \mathcal{L}(\mathbb{R}^3, \mathbb{R}^3)$ is given by $t(a, b, c) = (b, c, 0)$ then $\operatorname{Ker} t \subset \operatorname{Im} t$.

If t is a projection then $\operatorname{Im} t \cap \operatorname{Ker} t = \{0\}$ and none of the above are possible.

1.11 Consider the elements of $\mathcal{L}(\mathbb{R}^3, \mathbb{R}^3)$ given by

$$t_1(a, b, c) = (a, a, 0);$$
$$t_2(a, b, c) = (0, b, 0);$$
$$t_3(a, b, c) = (0, b, c);$$
$$t_4(a, b, c) = (0, b - a, c);$$
$$t_5(a, b, c) = (a, 0, 0).$$

Each of these transformations is a projection. We have

$$\operatorname{Ker} t_5 = \operatorname{Ker} t_1 \qquad \text{but} \qquad \operatorname{Im} t_5 \neq \operatorname{Im} t_1;$$
$$\operatorname{Im} t_3 = \operatorname{Im} t_4 \qquad \text{but} \qquad \operatorname{Ker} t_3 \neq \operatorname{Ker} t_4.$$

Also, $t_1 \circ t_2 = 0$ but $t_2 \circ t_1 \neq 0$. (Note that $t_2 \circ t_1$ is not a projection.)

1.12 Clearly, $e_1 + e_2$ is a projection if and only if (denoting composites by juxtaposition) $e_1 e_2 + e_2 e_1 = 0$. Thus if $e_1 e_2 = 0$ and $e_2 e_1 = 0$ then the property holds. Conversely, suppose that $e_1 + e_2$ is a projection. Then multiplying each side of $e_1 e_2 + e_2 e_1 = 0$ on the left by e_1 we obtain $e_1 e_2 + e_1 e_2 e_1 = 0$, and multiplying each side on the right by e_1 we obtain $e_1 e_2 e_1 + e_2 e_1 = 0$. It follows that $e_1 e_2 = e_2 e_1$. But $e_1 e_2 + e_2 e_1 = 0$ also gives $e_1 e_2 = -e_2 e_1$. Hence we have that each composite is zero.

When $e_1 + e_2$ is a projection, we have that

$$\operatorname{Ker}(e_1 + e_2) = \operatorname{Ker} e_1 \cap \operatorname{Ker} e_2,$$
$$\operatorname{Im}(e_1 + e_2) = \operatorname{Im} e_1 \oplus \operatorname{Im} e_2.$$

1.13 Take $U = \{(0, a, b) \mid a, b \in \mathbb{R}\}$. Then $\mathbb{R}^3 = V \oplus U$ since it is clear that $V \cap U = \{0\}$ and that

$$(a, b, c) = (a, a, 0) + (0, b - a, c).$$

For the last part refer to question 1.11; take $e = t_1$ and $f = t_4$.

1.14 Suppose that $\operatorname{Im} e = \operatorname{Im} f$. Then for every $v \in V$ we have $f(v) \in \operatorname{Im} f = \operatorname{Im} e$ so, since e acts as the identity map on its image, $e[f(v)] = f(v)$ and hence $e \circ f = f$. Likewise, $f \circ e = e$. Conversely, if $e \circ f = f$ and $f \circ e = e$, let $x \in \operatorname{Im} e$. Then $e(x) = x$ gives $e(x) = f[e(x)] = f(x) \in \operatorname{Im} f$ and so $\operatorname{Im} e \subseteq \operatorname{Im} f$. The reverse inclusion is obtained similarly.

Since $\operatorname{Im} e_1 = \cdots = \operatorname{Im} e_k$ we have $e_i \circ e_j = e_j$ for all i, j. Now

$$
\begin{aligned}
e^2 &= \left(\sum_{i=1}^{k} \lambda_i e_i \right)^2 \\
&= \lambda_1^2 e_1^2 + \cdots + \lambda_k^2 e_k^2 + \lambda_1 \lambda_2 e_1 e_2 + \lambda_2 \lambda_1 e_2 e_1 + \cdots + \lambda_k \lambda_{k-1} e_k e_{k-1} \\
&= \left(\sum_{i=1}^{k} \lambda_i \right) \lambda_1 e_1 + \cdots + \left(\sum_{i=1}^{k} \lambda_i \right) \lambda_k e_k \\
&= e,
\end{aligned}
$$

and so e is also a projection. To show that $\operatorname{Im} e = \operatorname{Im} e_1$ it suffices to prove that $e \circ e_1 = e_1$ and $e_1 \circ e = e$. Now

$$
(\lambda_1 e_1 + \cdots + \lambda_k e_k) e_1 = \lambda_1 e_1 + \cdots + \lambda_k e_1 = e_1
$$

gives the first of these, and the second is similar.

For the last part, consider $e, f \in \mathcal{L}(\mathbb{R}^2, \mathbb{R}^2)$ given by

$$
e(a, b) = (a, 0), \qquad f(a, b) = (0, b).
$$

Then e and f are projections but clearly $\frac{1}{2}e + \frac{1}{2}f$ is not.

1.15 Since sums and scalar multiples of step functions are step functions it is clear that E is a subspace of the real vector space of all mappings from $\mathbb{R}$ to $\mathbb{R}$. Given $\vartheta \in E$, the step function ϑ_i whose graph is

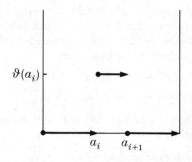

i.e. the function that agrees with ϑ on $[a_i, a_{i+1}[$ and is zero elsewhere, is given by the prescription

$$\vartheta_i(x) = \vartheta(a_i)[e_{a_i}(x) - e_{a_{i+1}}(x)].$$

It follows that $\{e_k \mid k \in [0,1[\}$ generates E since then

$$\vartheta = \sum_{i=0}^{n+1} \vartheta_i.$$

Since the functions e_k clearly form an independent set, they therefore form a basis of E.

It is likewise clear that F is a vector space and that G is a subspace of F. Consider now an element of F, as depicted in the diagram

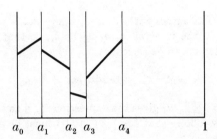

From geometric considerations it is clear that every element of F can be written uniquely as the sum of a function $e \in E$ and a function $g \in G$. [It helps to think of the above strips as pieces of wood that can slide up and down.] Thus it is clear that $F = E \oplus G$.

To show that $\{g_k \mid k \in [0,1[\}$ is a basis for G, observe first that the graph of $g_{a_i} - g_{a_{i+1}}$ is of the form

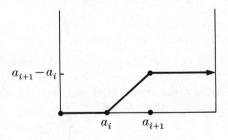

Given $\mu \in F$, consider the function μ_i whose graph is

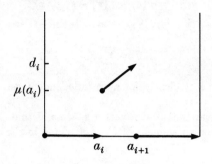

i.e. the function that agrees with μ on $[a_i, a_{i+1}[$ and is zero elsewhere. Let the gradient in the interval $[a_i, a_{i+1}[$ be b_i, so that $d_i = \mu(a_i) + b_i(a_{i+1} - a_i)$. Then it can be seen that

$$\mu_i = b_i(g_{a_i} - g_{a_{i+1}}) + \mu(a_i)e_{a_i} - d_i e_{a_{i+1}}.$$

Consequently $\mu = \sum_{i=0}^{n+1} \mu_i$ gives the expression for μ as a sum of $g \in G$ and $e \in E$. It now follows that $\{g_k \mid k \in [0, 1[\}$ must be a basis for G.

Finally, observe that

$$I(e_k) = \int_0^x e_k(t)\,dt = g_k$$

so that I carries a basis to a basis and therefore extends to an isomorphism from E to G.

1.16   Let $t_1, t_2 \in \mathcal{L}(\mathbb{R}^3, \mathbb{R}^3)$ be given by

$$t_1(a, b, c) = (a, 2b, 3c), \qquad t_2(a, b, c) = (2a, 3b, 6c).$$

Then t_1 has eigenvalues $1, 2, 3$ and t_2 has eigenvalues $2, 3, 6$. So both questions can be answered in the affirmative.

1.17   The eigenvalues of t are $1 + i\sqrt{2}$ and $1 - i\sqrt{2}$. Associated eigenvectors of any matrix representing t are $[1, i\sqrt{2}/2]$ and $[1, -i\sqrt{2}/2]$. Since the eigenvalues are distinct, the eigenvectors of t form a basis of $\mathbb{C}^2$. The matrix of t with respect to this basis is

$$\begin{bmatrix} 1 + i\sqrt{2} & 0 \\ 0 & 1 - i\sqrt{2} \end{bmatrix}.$$

1.18 Since t has 0 as an eigenvalue we have $t(v) = 0$ for some non-zero $v \in V$ and hence t is not injective, so not invertible. Thus if t is invertible then all its eigenvalues are non-zero. For the converse, suppose that t is not invertible and hence not injective. Then there is a non-zero vector $v \in \operatorname{Ker} t$, and $t(v) = 0$ shows that 0 is an eigenvalue of t.

If now t is invertible and $t(v) = \lambda v$ with $\lambda \neq 0$ then $v = t^{-1}[t(v)] = t^{-1}(\lambda v) = \lambda t^{-1}(v)$ gives $t^{-1}(v) = \lambda^{-1}v$ and so λ^{-1} is an eigenvalue of t^{-1} with the same associated eigenvector. (*Remark.* Note that we have assumed that V is finite-dimensional (where?)—in fact the result is false for infinite-dimensional spaces.)

1.19 Suppose that λ is a non-zero eigenvalue of t. Then $t(v) = \lambda v$ for some non-zero $v \in V$ and

$$0 = t^m(v) = t^{m-1}[t(v)] = t^{m-1}(\lambda v) = \cdots = \lambda^m v,$$

and we have the contradiction $\lambda^m = 0$. Hence all the eigenvalues of t are zero.

If t is diagonalisable then the matrix A of t is similar to the diagonal matrix

$$\begin{bmatrix} \lambda_1 & & \\ & \ddots & \\ & & \lambda_n \end{bmatrix}$$

where $\lambda_1, \ldots, \lambda_n$ are the eigenvalues of t. But we have just seen that all the eigenvalues are zero. Thus, for some invertible matrix P we have $P^{-1}AP = 0$ which gives $A = 0$ and hence the contradiction $t = 0$.

1.20 The matrix of t with respect to the canonical basis $\{(1,0),(0,1)\}$ is

$$\begin{bmatrix} 1 & 4 \\ \frac{1}{2} & -1 \end{bmatrix}.$$

The characteristic equation is $(\lambda - \sqrt{3})(\lambda + \sqrt{3}) = 0$ and, since $t - \sqrt{3}\,\mathrm{id} \neq 0, t + \sqrt{3}\,\mathrm{id} \neq 0$, the minimum polynomial is $(X - \sqrt{3})(X + \sqrt{3})$.

1.21 That t is linear follows from

$$t(f(X) + g(X)) = t((f + g)(X))$$
$$= (f + g)(X + 1)$$
$$= f(X + 1) + g(X + 1) = t(f(X)) + t(g(X)),$$

$$t(\lambda f(X)) = t((\lambda f)(X)) = \lambda f(X + 1) = \lambda t(f(X)).$$

The matrix of t relative to $\{1, X, \ldots, X^n\}$ is

$$\begin{bmatrix} 1 & 1 & 1 & 1 & \ldots & 1 \\ 0 & 1 & 2 & 3 & \ldots & n \\ 0 & 0 & 1 & 3 & \ldots & \frac{1}{2}n(n-1) \\ \vdots & \vdots & \vdots & \vdots & & \vdots \\ \vdots & & & & \ddots & \vdots \\ 0 & 0 & 0 & 0 & \ldots & 1 \end{bmatrix}.$$

The eigenvalues of t are all 1. The characteristic polynomial is

$$g(X) = (X - 1)^{n+1}.$$

Hence, by the Cayley–Hamilton theorem, $g(t) = 0$. The minimum polynomial of t is then $m(X) = (X - 1)^r$ for some r with $1 \le r \le n + 1$. A simple check using the above matrix shows that $(t - \mathrm{id}_V)^r \ne 0$ for $1 \le r \le n$. Consequently we have that $m(X) = (X - 1)^{n+1}$.

1.22   Let $\lambda_1, \ldots, \lambda_n$ be the eigenvalues of t. Then $\lambda_1^2, \ldots, \lambda_n^2$ are the eigenvalues of $t^2 = \mathrm{id}_V$ and so

$$\lambda_1^2 = \cdots = \lambda_n^2 = 1.$$

Consequently, $\lambda_i = \pm 1$ for each i and hence the sum of the eigenvalues of t is an integer.

1.23   (a) We have

$$\begin{vmatrix} 3 - \lambda & -1 \\ -1 & 3 - \lambda \end{vmatrix} = \lambda^2 - 6\lambda + 8 = (\lambda - 4)(\lambda - 2)$$

so the eigenvalues are 2 and 4, each of geometric multiplicity 1. For the eigenvectors associated with the eigenvalue 2, solve

$$\begin{bmatrix} 1 & -1 \\ -1 & 1 \end{bmatrix} \begin{bmatrix} x \\ y \end{bmatrix} = \begin{bmatrix} 0 \\ 0 \end{bmatrix}$$

to obtain the eigenspace $\{[x, x] \mid x \in \mathbb{R}\}$. For the eigenvectors associated with the eigenvalue 4, solve

$$\begin{bmatrix} -1 & -1 \\ -1 & -1 \end{bmatrix} \begin{bmatrix} x \\ y \end{bmatrix} = \begin{bmatrix} 0 \\ 0 \end{bmatrix}$$

to obtain the eigenspace $\{[x, -x] \mid x \in \mathbb{R}\}$. Since A has distinct eigenvalues it is diagonalisable. A suitable matrix P is

$$\begin{bmatrix} 1 & 1 \\ 1 & -1 \end{bmatrix}.$$

(b) The eigenvalues are 2, 3, 6 and are all of geometric multiplicity 1. Associated eigenvectors are $[1, 0, -1], [1, 1, 1], [1, -2, 1]$. A is diagonalisable; take, for example,

$$P = \begin{bmatrix} 1 & 1 & 1 \\ 0 & 1 & -2 \\ -1 & 1 & 1 \end{bmatrix}.$$

(c) The eigenvalues of A are 6, 6, and 12. For the eigenvalue 6 we consider

$$\begin{bmatrix} 1 & -1 & -2 \\ -1 & 1 & 2 \\ -2 & 2 & 4 \end{bmatrix} \begin{bmatrix} x \\ y \\ z \end{bmatrix} = \begin{bmatrix} 0 \\ 0 \\ 0 \end{bmatrix}.$$

We obtain $-x + y + 2z = 0$. We can therefore find two linearly independent eigenvectors associated with the eigenvalue 6, for example $[1, 1, 0]$ and $[2, 0, 1]$. Hence this eigenvalue has geometric multiplicity 2. The eigenvalue 12 has geometric multiplicity 1 and an associated eigenvector is $[-1, 1, 2]$. Hence A is diagonalisable and a suitable matrix P is

$$\begin{bmatrix} 1 & 2 & -1 \\ 1 & 0 & 1 \\ 0 & 1 & 2 \end{bmatrix}.$$

(d) The eigenvalues are 2, 2, 1. For the eigenvalue 2 we consider

$$\begin{bmatrix} 0 & 1 & -1 \\ 0 & 0 & 1 \\ 0 & 0 & -1 \end{bmatrix} \begin{bmatrix} x \\ y \\ z \end{bmatrix} = \begin{bmatrix} 0 \\ 0 \\ 0 \end{bmatrix}$$

from which we see that the corresponding eigenspace is spanned by $[1, 0, 0]$. Hence the eigenvalue 2 has geometric multiplicity 1. The eigenvalue 1 also has geometric multiplicity 1, the corresponding eigenspace being spanned by $[-2, 1, -1]$. In this case A is not diagonalisable.

(e) The eigenvalues are 2, 2, 2. From

$$\begin{bmatrix} -1 & 0 & 1 \\ 0 & 0 & 1 \\ -1 & 0 & 1 \end{bmatrix} \begin{bmatrix} x \\ y \\ z \end{bmatrix} = \begin{bmatrix} 0 \\ 0 \\ 0 \end{bmatrix}$$

we see that the corresponding eigenspace is spanned by $[0, 1, 0]$ and has dimension 1. Thus the geometric multiplicity of the eigenvalue 2 is 1, and A is not diagonalisable.

1.24 The matrix in question is

$$A = \begin{bmatrix} 1 & 2 \\ 1 & 1 \end{bmatrix},$$

and we have

$$\begin{bmatrix} a_n \\ b_n \end{bmatrix} = A^{n-1} \begin{bmatrix} 1 \\ 1 \end{bmatrix}.$$

The characteristic polynomial of A is $X^2 - 2X - 1$ and its eigenvalues are $\lambda_1 = 1 + \sqrt{2}$ and $\lambda_2 = 1 - \sqrt{2}$. Corresponding eigenvectors are $[\sqrt{2}, 1]$ and $[-\sqrt{2}, 1]$. The matrix

$$P = \begin{bmatrix} \sqrt{2} & -\sqrt{2} \\ 1 & 1 \end{bmatrix}$$

is such that $P^{-1}AP = \text{diag}\{\lambda_1, \lambda_2\}$.

In the new coordinate system, $\begin{bmatrix} 1 \\ 1 \end{bmatrix}$ becomes $P^{-1}\begin{bmatrix} 1 \\ 1 \end{bmatrix} = \begin{bmatrix} p_1 \\ p_2 \end{bmatrix}$ where

$$p_1 = \frac{\sqrt{2} + 1}{2\sqrt{2}}, \qquad p_2 = \frac{\sqrt{2} - 1}{2\sqrt{2}}.$$

We then have

$$a_n = p_1 \lambda_1^{n-1} \sqrt{2} - p_2 \lambda_2^{n-1} \sqrt{2}, \quad b_n = p_1 \lambda_1^{n-1} + p_2 \lambda_2^{n-1}$$

from which we see that

$$\begin{aligned}
\frac{a_n}{b_n} &= \frac{p_1 \lambda_1^{n-1} \sqrt{2} - p_2 \lambda_2^{n-1} \sqrt{2}}{p_1 \lambda_1^{n-1} + p_2 \lambda_2^{n-1}} \\
&= \frac{\sqrt{2} - (p_2/p_1)(\lambda_2/\lambda_1)^{n-1} \sqrt{2}}{1 + (p_2/p_1)(\lambda_2/\lambda_1)^{n-1}} \\
&= \sqrt{2} \left(\frac{1 - (p_2/p_1)(\lambda_2/\lambda_1)^{n-1}}{1 + (p_2/p_1)(\lambda_2/\lambda_1)^{n-1}} \right).
\end{aligned}$$

Now since $0 < |\lambda_2/\lambda_1| < 1$ we deduce that

$$\lim_{n \to \infty} \frac{a_n}{b_n} = \sqrt{2}.$$

1.25 Since $f(X)$ and $g(X)$ are coprime there are polynomials $p(X)$ and $q(X)$ such that $f(X)p(X) + g(X)q(X) = 1$. Let $c = p(t)$ and $d = q(t)$. Then $ac + bd = \mathrm{id}_V$.

Suppose now that v is an eigenvector of ab associated with the eigenvalue 0. (Note that ab has 0 as an eigenvalue since t is singular.) Let $u = a(cv)$ and $w = b(dv)$. Then since a, b, c commute we have

$$bu = bacv = cabv = 0,$$

and since a, b, d commute we have

$$aw = abdv = dabv = 0.$$

Also, $u + w = (ac + bd)v = v$ since $ac + bd = \mathrm{id}_V$.

1.26 If $u, v \in C_\lambda$ then $t(u) = \lambda u$ and $t(v) = \lambda v$ and so

$$t(au + bv) = at(u) + bt(v) = a\lambda u + b\lambda v = \lambda(au + bv)$$

and hence C_λ is a subspace of V.

Let $v \in C_\lambda$. Then, since s and t commute, we have

$$t[s(v)] = s[t(v)] = s(\lambda v) = \lambda s(v)$$

from which it follows that C_λ is s–invariant.

If λ_i is an eigenvalue then $C_{\lambda_i} \neq \{0\}$. If $\dim C_{\lambda_i} > 1$ then since t has n distinct eigenvalues this would give more than n linearly independent vectors, which is impossible. Hence C_{λ_i} is spanned by a single vector, v_i say. Since C_{λ_i} is s–invariant we have $s(v_i) = \mu_i v_i$ for some $\mu_i \in F$. Hence the matrix of s with respect to the basis $\{v_1, \ldots, v_n\}$ is diagonal.

1.27 There is an integer $r \leq n$ with $\{u, t(u), \ldots, t^{r-1}(u)\}$ linearly independent and $\{u, t(u), \ldots, t^r(u)\}$ linearly dependent. Then

$$t^r(u) = a_0 u + a_1 t(u) + \cdots + a_{r-1} t^{r-1}(u),$$

so that, applying t,

$$t^{r+1}(u) = a_0 t(u) + a_1 t^2(u) + \cdots + a_{r-1} t^r(u).$$

Continuing in this way we see that $\{u, t(u), \ldots, t^{r-1}(u)\}$ spans U and so is a basis. By the above argument we have $t[t^i(u)] = \sum_{j=1}^{i} a_j t^j(u)$, so U is t–invariant.

Let $f(X) = -a_0 - a_1 X - \cdots - a_{r-1} X^{r-1} + X^r$. Then $f(X)$ has degree r and $[f(t)](u) = 0$. Since U is t–invariant the restriction t_U of t to U induces a linear transformation from U to itself. Now $[f(t_U)](u) = 0$ so $[f(t_U)](v) = 0$ for all $v \in U$. Hence $f(t_U)$ is the zero transformation on U. If now $g(X)$ is a polynomial of degree less than r with $g(t_U) = 0$ then $[g(t_U)](u) = 0$ implies that $\{u, t(u), \ldots, t^{r-1}(u)\}$ is dependent. Hence $f(X)$ is the (monic) polynomial of least degree such that $f(t_U) = 0$, so $f(X)$ is the minimum polynomial of t_U.

When $u = (1, 1, 0) \in \mathbb{R}^3$ and $t(x, y, z) = (x + y, x - y, z)$ we have that $U = \{(1, 1, 0), (2, 0, 0)\}$. Also, $t^2(u) = (2, 2, 0) = 2u$ and so $f(X) = X^2 - 2$.

1.28   (1) Proceed by induction. The result clearly holds for $n = 1$. In this example it is necessary, in order to apply the second principle of induction, to include a proof for $n = 2$:

$$
\begin{aligned}
q^2 &= r(s + t)r(s + t) \\
&= (rs + rt)^2 \\
&= rsrs + rtrs + rsrt + rtrt \\
&= rsr(s + t) \\
&= pq.
\end{aligned}
$$

Suppose now that it is true for all $r < n$ where $n > 2$. Then

$$q^{n+1} = q^n.q = p^{n-1}q^2 = p^{n-1}pq = p^n q,$$

which shows that it holds for $n + 1$. The second equality is established in a similar way.

(2) If r is non-singular then r^{-1} exists and consequently from $rtr = 0$ we obtain the contradiction $t = 0$. Hence r is singular, so both p and q are singular and hence have 0 as an eigenvalue. Consequently we see that $p(X)$ and $q(X)$ are divisible by X.

(3) Let $q(X) = a_1 X + a_2 X^2 + \cdots + a_r X^r$. Then $a_1 q + a_2 q^2 + \cdots + a_r q^r = 0$ and so $(a_1 q + \cdots + a_r q^r)p = 0$, i.e. $a_1 p^2 + \cdots + a_r p^{r+1} = 0$ which shows that p satisfies $Xq(X) = 0$. Similarly, q satisfies $Xp(X) = 0$.

By (3), $p(X)$ divides $Xq(X)$, and $q(X)$ divides $Xp(X)$, so we have

$$Xq(X) = p_1(X)p(X), \qquad Xp(X) = q_1(X)q(X)$$

for monic polynomials $p_1(X), q_1(X)$. Now $X^2 q(X) = X p_1(X) p(X) = p_1(X) q_1(X) q(X)$ so, since $q(X) \neq 0$, we have $p_1(X) q_1(X) = X^2$. Consequently, either (i) $p_1(X) = q_1(X) = X$, or (ii) $q_1(X) = X^2$, or (iii) $p_1(X) = X^2$.

1.29 Clearly, adding together the elements in the middle row, the middle column, and both diagonals, we obtain

$$\sum_{i,j} m_{ij} + 3m_{22} = 4\vartheta,$$

so that $3\vartheta + 3m_{22} = 4\vartheta$ and hence $\vartheta = 3m_{22}$.

If $m_{22} = \alpha, m_{11} = \alpha + \beta$ and $m_{31} = \alpha + \gamma$ then

$$m_{21} = 3\alpha - m_{11} - m_{31} = 3\alpha - \alpha - \beta - \alpha - \gamma = \alpha - \beta - \gamma,$$
$$m_{23} = 3\alpha - m_{21} - m_{22} = 3\alpha - \alpha + \beta + \gamma - \alpha = \alpha + \beta + \gamma,$$

and so on, and we obtain

$$M(\alpha,\beta,\gamma) = \begin{bmatrix} \alpha+\beta & \alpha-\beta+\gamma & \alpha-\gamma \\ \alpha-\beta-\gamma & \alpha & \alpha+\beta+\gamma \\ \alpha+\gamma & \alpha+\beta-\gamma & \alpha-\beta \end{bmatrix}.$$

It is readily seen that sums and scalar multiples of magic matrices are also magic. Hence the magic matrices constitute a subspace of $\mathrm{Mat}_{3\times3}(\mathbb{C})$. Also,

$$M(\alpha,\beta,\gamma) = \alpha M(1,0,0) + \beta M(0,1,0) + \gamma M(0,0,1)$$

so that B generates this subspace. Since $M(\alpha,\beta,\gamma) = 0$ if and only if $\alpha = \beta = \gamma = 0$, it follows that B is a basis for this subspace.

That $e_1 + e_2 + e_3$ is an eigenvector of f follows from the fact that

$$M(\alpha,\beta,\gamma)\begin{bmatrix}1\\1\\1\end{bmatrix} = \begin{bmatrix}3\alpha\\3\alpha\\3\alpha\end{bmatrix} = 3\alpha\begin{bmatrix}1\\1\\1\end{bmatrix}.$$

To compute the matrix of f relative to the basis $\{e_1 + e_2 + e_3, e_2, e_3\}$ we observe that, by the above,

$$f(e_1 + e_2 + e_3) = 3\alpha(e_1 + e_2 + e_3);$$

49

and that

$$f(e_2) = (\alpha - \beta + \gamma)e_1 + \alpha e_2 + (\alpha + \beta - \gamma)e_3$$
$$= (\alpha - \beta + \gamma)(e_1 + e_2 + e_3 - e_2 - e_3) + \alpha e_2 + (\alpha + \beta - \gamma)e_3$$
$$= (\alpha - \beta + \gamma)(e_1 + e_2 + e_3) + (\beta - \gamma)e_2 + (2\beta - 2\gamma)e_3,$$
$$f(e_3) = (\alpha - \gamma)e_1 + (\alpha + \beta + \gamma)e_2 + (\alpha - \beta)e_3$$
$$= (\alpha - \gamma)(e_1 + e_2 + e_3 - e_2 - e_3) + (\beta + 2\gamma)e_2 + (\gamma - \beta)e_3$$
$$= (\alpha - \gamma)(e_1 + e_2 + e_3) + (\beta + 2\gamma)e_2 + (\gamma - \beta)e_3.$$

The matrix of f relative to $\{e_1 + e_2 + e_3, e_2, e_3\}$ is then

$$L = \begin{bmatrix} 3\alpha & \alpha - \beta + \gamma & \alpha - \gamma \\ 0 & \beta - \gamma & \beta + 2\gamma \\ 0 & 2\beta - 2\gamma & \gamma - \beta \end{bmatrix}.$$

Since L and $M(\alpha, \beta, \gamma)$ represent the same linear mapping they are similar and therefore have the same eigenvalues. It is readily seen that

$$\det(L - \lambda I_3) = (3\alpha - \lambda)(\lambda^2 - 3\beta^2 + 3\gamma^2),$$

so the eigenvalues are 3α and $\pm\sqrt{3(\beta^2 - \gamma^2)}$.

1.30 If f is nilpotent of index p then $f^p = 0$ and $f^{p-1} \neq 0$. Let $x \in V$ be such that $f^{p-1}(x) \neq 0$ and consider the set

$$B_p = \{x, f(x), \ldots, f^{p-1}(x)\}.$$

Suppose that

$$(\star) \qquad \lambda_0 x + \lambda_1 f(x) + \cdots + \lambda_{p-1} f^{p-1}(x) = 0.$$

On applying f^{p-1} to $(\star)$ and using the fact that $f^p = 0$, we see that $\lambda_0 f^{p-1}(x) = 0$ whence we deduce that $\lambda_0 = 0$. Deleting the first term in $(\star)$ and applying f^{p-2} to the remainder, we obtain similarly $\lambda_1 = 0$. Repeating this argument, we see that each $\lambda_i = 0$ and hence that B_p is linearly independent.

It follows from the above that if f is nilpotent of index $n = \dim V$ then

$$B_n = \{x, f(x), \ldots, f^{n-1}(x)\}$$

is a basis of V. The matrix of f relative to B_n is readily seen to be

$$I_\star = \begin{bmatrix} 0 & 0 \\ I_{n-1} & 0 \end{bmatrix}.$$

Consequently, if A is an $n \times n$ matrix over F that is nilpotent of index n then A is similar to $I_\star$. Conversely, if A is similar to $I_\star$ then there is an invertible matrix P such that $P^{-1}AP = I_\star$, so that $A = PI_\star P^{-1}$. Computing the powers of A we see that

(i) $A^n = 0$;

(ii) $[A^{n-1}]_{n1} = 1$, so $A^{n-1} \neq 0$.

Hence A is nilpotent of index n.

1.31 To see that V is a $\mathbb{C}$–vector space it suffices to check the axioms concerning the external law. For example,

$$\begin{aligned}
(\alpha + i\beta)[(\gamma + i\delta)x] &= (\alpha + i\beta)[\gamma x - \delta f(x)] \\
&= \alpha[\gamma x - \delta f(x)] - \beta f[\gamma x - \delta f(x)] \\
&= \alpha\gamma x - \alpha\delta f(x) - \beta\gamma f(x) - \beta\delta x \\
&= (\alpha\gamma - \beta\delta)x - (\alpha\delta + \beta\gamma)f(x) \\
&= [(\alpha + i\beta)(\gamma + i\delta)]x.
\end{aligned}$$

Suppose now that $\{v_1, \ldots, v_r\}$ is a linearly independent subset of the $\mathbb{C}$–vector space V and that in the $\mathbb{R}$–vector space V we have

$$\sum \alpha_j v_j + \sum \beta_j f(v_j) = 0.$$

Using the given identity, we can rewrite this as the following equation in the $\mathbb{C}$–vector space V :

$$\sum (\alpha_j - i\beta_j)v_j = 0.$$

It follows that $\alpha_j - i\beta_j = 0$ for every j, so that $\alpha_j = 0 = \beta_j$. Consequently,

$$\{v_1, \ldots, v_r, f(v_1), \ldots, f(v_r)\}$$

is linearly independent in the $\mathbb{R}$–vector space V. Since V is of finite dimension over $\mathbb{R}$ it must then be so over $\mathbb{C}$. The given identity shows that every complex linear combination of $\{v_1, \ldots, v_n\}$ can be written as a real linear combination of

$$v_1, \ldots, v_n, f(v_1), \ldots, f(v_n).$$

If $\dim_{\mathbb{C}} V = n$ it then follows that $\dim_{\mathbb{R}} V = 2n$.

By considering a basis of V (over $\mathbb{R}$) of the form

$$\{v_1, \ldots, v_n, f(v_1), \ldots, f(v_n)\}$$

we deduce immediately from the fact that $f \circ f = -\operatorname{id}_V$ that the matrix of f relative to this basis is

$$\Gamma = \begin{bmatrix} 0 & -I_n \\ I_n & 0 \end{bmatrix}.$$

Clearly, it follows from the above that if A is a $2n \times 2n$ matrix over $\mathbb{R}$ such that $A^2 = -I_{2n}$ then A is similar to Γ. Conversely, if A is similar to Γ then there is an invertible matrix P such that $P^{-1}AP = \Gamma$ and hence $A^2 = (P\Gamma P^{-1})^2 = P\Gamma^2 P^{-1} = P(-I_{2n})P^{-1} = -I_{2n}$.

1.32 Let x be an eigenvector corresponding to λ. Then from $Ax = \lambda x$ we have that $x^t A^t = \lambda x^t$ and hence $\overline{x}^t \overline{A}^t = \overline{\lambda}\overline{x}^t$. Since $\overline{A} = A$ and $A^t = -A$ we deduce that $-\overline{x}^t A = \overline{\lambda}\overline{x}^t$. Thus $\overline{x}^t A x = -\overline{\lambda}\overline{x}^t x$. But we also have $\overline{x}^t A x = \overline{x}^t \lambda x = \lambda \overline{x}^t x$. It follows that $\lambda = -\overline{\lambda}$, so the real part of λ is zero. We also deduce from $Ax = \lambda x$ that $\overline{A}\overline{x} = \overline{\lambda}\overline{x}$, i.e. that $A\overline{x} = \overline{\lambda}\overline{x}$, so $\overline{\lambda}$ is also an eigenvalue.

$Y = (A - \lambda I)Z$ gives $Y^t = Z^t(A^t - \lambda I) = -Z^t(A + \lambda I)$ and hence $\overline{Y}^t = -\overline{Z}^t(\overline{A} + \overline{\lambda}I) = -\overline{Z}^t(A - \lambda I)$. Consequently,

$$\overline{Y}^t Y = -\overline{Z}^t(A - \lambda I).(A - \lambda I)Z = 0$$

since it is given that $(A - \lambda I)^2 Z = 0$. Now the elements of $\overline{Y}^t Y$ are of the form

$$\begin{bmatrix} a - ib & \ldots & x - iy \end{bmatrix} \begin{bmatrix} a + ib \\ \vdots \\ x + iy \end{bmatrix} = a^2 + b^2 + \cdots + x^2 + y^2$$

and a sum of squares is zero if and only if each summand is zero. Hence we see that $Y = 0$.

The minimum polynomial of A cannot have repeated roots. For, if this were of the form $m(X) = (X - \alpha)^2 p(X)$ then from $(A - \alpha I)^2 p(A) = 0$ we would have, by the above applied to each column of $p(A)$ in turn, $(A - \alpha I)p(A) = 0$ and $m(X)$ would not be the minimum polynomial.

Thus the minimum polynomial has simple roots and so A is similar to a diagonal matrix.

Suppose now that $Ax = i\alpha x$. Then $A\bar{x} = -i\alpha\bar{x}$ and

$$Au = A(x + \bar{x}) = i\alpha x - i\alpha\bar{x} = i\alpha(x - \bar{x}) = \alpha v,$$
$$Av = Ai(x - \bar{x}) = -\alpha x - \alpha\bar{x} = -\alpha(x + \bar{x}) = -\alpha u.$$

These equalities can be written in the form

$$\begin{bmatrix} Au \\ Av \end{bmatrix} = \begin{bmatrix} 0 & \alpha \\ -\alpha & 0 \end{bmatrix} \begin{bmatrix} u \\ v \end{bmatrix}.$$

The last part follows by choosing $i\alpha_1, \ldots, i\alpha_k$ to be the non-zero eigenvalues of A.

1.33 Since t satisfies its characteristic equation we have

$$(t - \mathrm{id})(t + 2\,\mathrm{id})(t - 2\,\mathrm{id}) = 0,$$

which gives $t^3 = t^2 + 4t - 4\,\mathrm{id}$. It is now readily seen that

$$t^4 = t^2 + 4(t^2 - \mathrm{id});$$
$$t^6 = t^2 + 4(1 + 4)(t^2 - \mathrm{id});$$
$$t^8 = t^2 + 4(1 + 4 + 4^2)(t^2 - \mathrm{id}).$$

This suggests that in general

$$t^{2p} = t^2 + 4(1 + 4 + 4^2 + \cdots + 4^{p-2})(t^2 - \mathrm{id}).$$

It is easy to see by induction that this is indeed the case. Thus we see that

$$t^{2n} = t^2 + 4(1 + 4 + \cdots + 4^{n-2})(t^2 - \mathrm{id})$$
$$= t^2 + 4\left(\frac{1 - 4^{n-1}}{1 - 4}\right)(t^2 - \mathrm{id})$$
$$= t^2 + \tfrac{4}{3}(4^{n-1} - 1)(t^2 - \mathrm{id}).$$

1.34 We have that

$$f(b_1) = \lambda_1 b_1;$$
$$(i \geq 2) \quad f(b_i') = \beta'_{1i} b_1 + \sum_{j \geq 2} m_{ji} b_j'.$$

Thus the matrix of f relative to the basis $\{b_1, b'_2, \ldots, b'_m\}$ is of the form

$$A = \begin{bmatrix} \lambda_1 & \beta'_{12} & \cdots & \beta'_{1n} \\ & & & \\ 0 & & M & \end{bmatrix}.$$

If $w \in W$, say $w = w_1 b_1 + \sum_{i \geq 2} w_i b'_i$ then

$$g(w) = \pi[f(w)] = \pi\left(w_1 \lambda_1 b_1 + \sum_{i \geq 2} w_i f(b'_i)\right)$$

$$= \sum_{i \geq 2} w_i f(b'_i) \in W,$$

since $b_1 \in \text{Ker}\,\pi$ and π acts as the identity on $\text{Im}\,\pi = W$. Thus W is g-invariant.

It is clear that $\text{Mat}\,(g', (b'_i)) = M$. Also, the characteristic equation of f is given by $\det(A - XI_n) = 0$, i.e. by

$$(\lambda_1 - X)\det(M - XI_n) = 0.$$

So the eigenvalues of g' are precisely those of f with the algebraic multiplicity of λ_1 reduced by 1. Since all the eigenvalues of f belong to F by hypothesis, so then do all those of g'.

The last part follows from the above by a simple inductive argument; if the result holds for $(n-1) \times (n-1)$ matrices then it holds for M and hence for A.

1.35   The eigenvalues of t are 0, 1, 1. The minimum polynomial is either $X(X-1)$ or $X(X-1)^2$. But $t^2 - t \neq 0$ so the minimum polynomial is $X(X-1)^2$. We have that

$$V = \text{Ker}\,t \oplus \text{Ker}(t - \text{id}_V)^2.$$

We must find a basis $\{w_1, w_2, w_3\}$ with

$$t(w_1) = 0, \quad (t - \text{id}_V)(w_2) = 0, \quad (t - \text{id}_V)(w_3) = \lambda w_2.$$

A suitable basis is $\{(-1, 2, 0), (1, -1, 0), (1, 1, 1)\}$, with respect to which the matrix of t is

$$\begin{bmatrix} 0 & 0 & 0 \\ 0 & 1 & 1 \\ 0 & 0 & 1 \end{bmatrix}.$$

1.36 We have that

$$t(1) = -5 - 8X - 5X^2,$$
$$t^2(1) = -5(-5 - 8X - 5X^2) - 8(1 + X + X^2) - 5(4 + 7X + 4X^2),$$
$$t^3(1) = 0.$$

Similarly we have that $t^3(X) = 0$ and $t^3(X^2) = 0$. Consequently $t^3 = 0$ and so t is nilpotent.

Take $v_1 = 1 + X + X^2$. Then we have $t(v_1) = 0$. Now take $v_2 = 5 + 8X + 5X^2$. Then we have

$$t(v_2) = 3(1 + X + X^2) \in \operatorname{span} \{v_1\}.$$

Finally, take $v_3 = 1$ and observe that

$$t(1) = -5 - 8X - 5X^2 \in \operatorname{span} \{v_1, v_2\}.$$

It is now clear that $\{1 + X + X^2, 5 + 8X + 5X^2, 1\}$ is a basis with respect to which the matrix of t is upper triangular.

1.37 (a) The characteristic polynomial is $X^2 + 2X + 1$ so the eigenvalues are -1 (twice). The corresponding eigenvector satisfies

$$\begin{bmatrix} 40 & -64 \\ 25 & -40 \end{bmatrix} \begin{bmatrix} x \\ y \end{bmatrix} = \begin{bmatrix} 0 \\ 0 \end{bmatrix}$$

so -1 has geometric multiplicity 1 with $[8, 5]$ as an associated eigenvector. Hence the Jordan normal form is

$$\begin{bmatrix} -1 & 1 \\ 0 & -1 \end{bmatrix}.$$

A Jordan basis can be found by solving

$$(A + I_2)v_1 = 0, \quad (A + I_2)v_2 = v_1.$$

Take $v_1 = [8, 5]$. Then a possible solution for v_2 is $[5, 3]$, giving

$$P = \begin{bmatrix} 8 & 5 \\ 5 & 3 \end{bmatrix}.$$

(b) The characteristic polynomial is $(X+1)^2$. The eigenvalues are -1 (twice) with geometric multiplicity 1, and a corresponding eigenvector is $[1,0]$. The Jordan normal form is

$$\begin{bmatrix} -1 & 1 \\ 0 & -1 \end{bmatrix}.$$

A Jordan basis satisfies

$$(A + I_2)v_1 = 0, \quad (A + I_2)v_2 = v_1.$$

Take $v_1 = [1,0]$ and $v_2 = [0,-1]$; then

$$P = \begin{bmatrix} 1 & 0 \\ 0 & -1 \end{bmatrix}.$$

(Any Jordan basis is of the form $\{[c,0],[d,-c]\}$ with $P = \begin{bmatrix} c & d \\ 0 & -c \end{bmatrix}$.)

(c) The characteristic polynomial is $(X - 1)^3$, so the only eigenvalue is 1. It has geometric multiplicity 2 with $\{[1,0,0],[0,2,3]\}$ as a basis for the eigenspace. The Jordan normal form is then

$$\begin{bmatrix} 1 & 1 & 0 \\ 0 & 1 & 0 \\ 0 & 0 & 1 \end{bmatrix}.$$

A Jordan basis satisfies

$$(A - I_3)v_1 = 0, \quad (A - I_3)v_2 = v_1, \quad (A - I_3)v_3 = 0.$$

Now $(A - I_3)^2 = 0$ so choose v_2 to be any vector not in $\langle [1,0,0],[0,2,3] \rangle$, for example $v_2 = [0,1,0]$. Then $v_1 = (A - I_3)v_2 = [3,6,9]$. For v_3 choose any vector in $\langle [1,0,0],[0,2,3] \rangle$ that is independent of $[3,6,9]$, for example $v_3 = [1,0,0]$. This gives

$$P = \begin{bmatrix} 3 & 0 & 1 \\ 6 & 1 & 0 \\ 9 & 0 & 0 \end{bmatrix}.$$

(d) The Jordan normal form is

$$\begin{bmatrix} 3 & 1 & 0 \\ 0 & 3 & 0 \\ 0 & 0 & 3 \end{bmatrix}.$$

A Jordan basis satisfies

$$(A - 3I_3)v_1 = 0, \quad (A - 3I_3)v_2 = v_1, \quad (A - 3I_3)v_3 = 0.$$

Choose $v_2 = [0, 0, 1]$. Then $v_1 = [1, 0, 0]$ and a suitable choice for v_3 is $[0, 1, 0]$. Thus

$$P = \begin{bmatrix} 1 & 0 & 0 \\ 0 & 0 & 1 \\ 0 & 1 & 0 \end{bmatrix}.$$

1.38 The characteristic polynomial of A is $(X-1)^3(X-2)^2$. For the eigenvalue 2 we solve

$$\begin{bmatrix} 0 & 1 & 1 & 1 & 0 \\ 0 & 0 & 0 & 0 & 0 \\ 0 & 0 & 0 & 1 & 0 \\ 0 & 0 & 0 & -1 & 1 \\ 0 & -1 & -1 & -1 & -2 \end{bmatrix} \begin{bmatrix} x \\ y \\ z \\ t \\ w \end{bmatrix} = \begin{bmatrix} 0 \\ 0 \\ 0 \\ 0 \\ 0 \end{bmatrix}$$

to obtain $w = t = 0, y + z = 0$. Thus the general eigenvector associated with the eigenvalue 2 is $[x, y, -y, 0, 0]$ with x, y not both zero. The Jordan block associated with the eigenvalue 2 is

$$\begin{bmatrix} 2 & 0 \\ 0 & 2 \end{bmatrix}.$$

For the eigenvalue 1 we solve

$$\begin{bmatrix} 1 & 1 & 1 & 1 & 0 \\ 0 & 1 & 0 & 0 & 0 \\ 0 & 0 & 1 & 1 & 0 \\ 0 & 0 & 0 & 0 & 1 \\ 0 & -1 & -1 & -1 & -1 \end{bmatrix} \begin{bmatrix} x \\ y \\ z \\ t \\ w \end{bmatrix} = \begin{bmatrix} 0 \\ 0 \\ 0 \\ 0 \\ 0 \end{bmatrix}$$

to obtain $w = y = x = 0, z + t = 0$. Thus the general eigenvector associated with the eigenvalue 1 is $[0, 0, z, -z, 0]$ with $z \neq 0$. The Jordan block associated with the eigenvalue 1 is

$$\begin{bmatrix} 1 & 1 & 0 \\ 0 & 1 & 1 \\ 0 & 0 & 1 \end{bmatrix}.$$

The Jordan normal form of A is therefore

$$\begin{bmatrix} 2 & 0 & 0 & 0 & 0 \\ 0 & 2 & 0 & 0 & 0 \\ 0 & 0 & 1 & 1 & 0 \\ 0 & 0 & 0 & 1 & 1 \\ 0 & 0 & 0 & 0 & 1 \end{bmatrix}.$$

Take $[0, 0, 1, -1, 0]$ as an eigenvector associated with the eigenvalue 1. Then we solve

$$\begin{bmatrix} 1 & 1 & 1 & 1 & 0 \\ 0 & 1 & 0 & 0 & 0 \\ 0 & 0 & 1 & 1 & 0 \\ 0 & 0 & 0 & 0 & 1 \\ 0 & -1 & -1 & -1 & -1 \end{bmatrix} \begin{bmatrix} x \\ y \\ z \\ t \\ w \end{bmatrix} = \begin{bmatrix} 0 \\ 0 \\ 1 \\ -1 \\ 0 \end{bmatrix}$$

to obtain $y = 0, w = -1, z + t = 1, x = -1$, so we take $[-1, 0, 0, 1, -1]$. Next we solve

$$\begin{bmatrix} 1 & 1 & 1 & 1 & 0 \\ 0 & 1 & 0 & 0 & 0 \\ 0 & 0 & 1 & 1 & 0 \\ 0 & 0 & 0 & 0 & 1 \\ 0 & -1 & -1 & -1 & -1 \end{bmatrix} \begin{bmatrix} x \\ y \\ z \\ t \\ w \end{bmatrix} = \begin{bmatrix} -1 \\ 0 \\ 0 \\ 1 \\ -1 \end{bmatrix}$$

to obtain $y = 0, t + z = 0, w = 1, x = -1$, so we consider $[-1, 0, 0, 0, 1]$.

A Jordan basis is therefore

$$\{[1, 0, 0, 0, 0], [0, 1, -1, 0, 0], [0, 0, 1, -1, 0], [-1, 0, 0, 1, -1], [-1, 0, 0, 0, 1]\}$$

and a suitable matrix is

$$P = \begin{bmatrix} 1 & 0 & 0 & -1 & -1 \\ 0 & 1 & 0 & 0 & 0 \\ 0 & -1 & 1 & 0 & 0 \\ 0 & 0 & -1 & 1 & 0 \\ 0 & 0 & 0 & -1 & 1 \end{bmatrix}.$$

1.39 (a) The Jordan form and a suitable (non-unique) matrix P are

$$J = \begin{bmatrix} 2 & 1 & 0 \\ 0 & 2 & 0 \\ 0 & 0 & 2 \end{bmatrix}, \quad P = \begin{bmatrix} 2 & -5 & 5 \\ 2 & -3 & 8 \\ 3 & 8 & -7 \end{bmatrix}.$$

(b) The Jordan form and a suitable P are

$$\begin{bmatrix} 2 & 0 & 0 & 0 \\ 0 & 1 & 1 & 0 \\ 0 & 0 & 1 & 1 \\ 0 & 0 & 0 & 1 \end{bmatrix}, \quad P = \begin{bmatrix} 4 & 3 & 2 & 0 \\ 5 & 4 & 3 & 0 \\ -2 & -2 & -1 & 0 \\ 11 & 6 & 4 & 1 \end{bmatrix}.$$

1.40 The Jordan normal form is

$$\begin{bmatrix} 2 & 0 & 0 & 0 & 0 \\ 0 & 2 & 1 & 0 & 0 \\ 0 & 0 & 2 & 0 & 0 \\ 0 & 0 & 0 & 3 & 1 \\ 0 & 0 & 0 & 0 & 3 \end{bmatrix}.$$

A Jordan basis is

$$\{[2,1,0,0,1],[1,0,1,0,0],[0,1,0,1,0],[-1,0,0,1,0],[2,0,0,0,1]\}.$$

1.41 The minimum polynomial is $(X-2)^3$. There are two possibilities for the Jordan normal form, namely

$$J_1 = \begin{bmatrix} 2 & 1 & 0 & 0 & 0 \\ 0 & 2 & 1 & 0 & 0 \\ 0 & 0 & 2 & 0 & 0 \\ 0 & 0 & 0 & 2 & 0 \\ 0 & 0 & 0 & 0 & 2 \end{bmatrix}, \quad J_2 = \begin{bmatrix} 2 & 1 & 0 & 0 & 0 \\ 0 & 2 & 0 & 0 & 0 \\ 0 & 0 & 2 & 1 & 0 \\ 0 & 0 & 0 & 2 & 1 \\ 0 & 0 & 0 & 0 & 2 \end{bmatrix}.$$

Each of these has $(X-2)^3$ as minimum polynomial. There are two linearly independent eigenvectors; e.g., $[0,-1,1,1,0]$ and $[0,1,0,0,1]$. The number of linearly independent eigenvectors does not determine the Jordan form. For example, the matrix J_2 above and the matrix

$$\begin{bmatrix} 2 & 0 & 0 & 0 & 0 \\ 0 & 2 & 1 & 0 & 0 \\ 0 & 0 & 2 & 1 & 0 \\ 0 & 0 & 0 & 2 & 1 \\ 0 & 0 & 0 & 0 & 2 \end{bmatrix}$$

have two linearly independent eigenvectors. Both pieces of information are required in order to determine the Jordan form. For the given matrix this is J_2.

1.42   A basis for $\mathbb{R}_4[X]$ is $\{1, X, X^2, X^3\}$ and $D(1) = 0, D(X) = 1, D(X^2) = 2X, D(X^3) = 3X^2$. Hence, relative to the above basis, D is represented by the matrix

$$\begin{bmatrix} 0 & 1 & 0 & 0 \\ 0 & 0 & 2 & 0 \\ 0 & 0 & 0 & 3 \\ 0 & 0 & 0 & 0 \end{bmatrix}.$$

The characteristic polynomial of this matrix is X^4, the only (quadruple) eigenvalue is 0, and the eigenspace of 0 is of dimension 1 with basis $\{1\}$. So the Jordan normal form is

$$\begin{bmatrix} 0 & 1 & 0 & 0 \\ 0 & 0 & 1 & 0 \\ 0 & 0 & 0 & 1 \\ 0 & 0 & 0 & 0 \end{bmatrix}.$$

A Jordan basis is $\{f_1, f_2, f_3, f_4\}$ where

$$Df_1 = 0, \ Df_2 = f_1, \ Df_3 = f_2, \ Df_4 = f_3.$$

Choose $f_1 = 1$; then $f_2 = X, f_3 = \frac{1}{2}X^2, f_4 = \frac{1}{6}X^3$ so a Jordan basis is $\{6, 6X, 3X^2, X^3\}$.

1.43   The possible Jordan forms are

$$\begin{bmatrix} 3 & 0 & 0 \\ 0 & 3 & 0 \\ 0 & 0 & 3 \end{bmatrix}, \begin{bmatrix} 3 & 1 & 0 \\ 0 & 3 & 1 \\ 0 & 0 & 3 \end{bmatrix}, \begin{bmatrix} 3 & 1 & 0 \\ 0 & 3 & 0 \\ 0 & 0 & 3 \end{bmatrix}, \begin{bmatrix} 3 & 0 & 0 \\ 0 & 3 & 1 \\ 0 & 0 & 3 \end{bmatrix}.$$

The last two are similar.

1.44   (i) and (ii) are true : use the fact that AB and $BA = A^{-1}(AB)A$ are similar.

(iii) and (iv) are false; for example,

$$\begin{bmatrix} 0 & 0 \\ 0 & 0 \end{bmatrix}\begin{bmatrix} 1 & 0 \\ 0 & 0 \end{bmatrix} \text{ and } \begin{bmatrix} 1 & 0 \\ 0 & 0 \end{bmatrix}\begin{bmatrix} 0 & 0 \\ 0 & 0 \end{bmatrix}$$

clearly have the same Jordan normal form.

1.45 V decomposes into a direct sum of t–invariant subspaces, say $V = V_1 \oplus \cdots \oplus V_r$, and each summand is associated with one and only one eigenvalue of t. Without loss of generality we can assume that t has a single eigenvalue λ. Consider an $i \times i$ Jordan block. Corresponding to this block there are i basis elements of V, say $v_1, \ldots, v_i$, with

$$(t - \lambda \operatorname{id}_V)v_1 = 0;$$
$$(t - \lambda \operatorname{id}_V)^2 v_2 = (t - \lambda \operatorname{id}_V)v_1 = 0;$$
$$\vdots$$
$$(t - \lambda \operatorname{id}_V)^i v_i = (t - \lambda \operatorname{id}_V)^{i-1} v_{i-1} = \cdots = 0.$$

Thus there is one eigenvector associated with each block, and so there are

$$\dim \operatorname{Ker}(t - \lambda \operatorname{id}_V)$$

blocks.

Consider $\operatorname{Ker}(t - \lambda \operatorname{id}_V)^j$. For every 1×1 block there corresponds a single basis element which is an eigenvector in $\operatorname{Ker}(t - \lambda \operatorname{id}_V)^j$. For every 2×2 block there correspond two basis elements in $\operatorname{Ker}(t - \lambda \operatorname{id}_V)^j$ if $j \geq 2$ and 1 basis element if $j < 2$. In general, to each $i \times i$ block there correspond i basis elements in $\operatorname{Ker}(t - \lambda \operatorname{id}_V)^j$ if $j \geq i$ and j basis elements if $j < i$.

It follows that

$$d_j = n_1 + 2n_2 + \cdots + (j-1)n_{j-1} + j(n_j + n_{j+1} + \cdots)$$

and a simple calculation shows that $2d_i - d_{i-1} - d_{i+1} = n_i$.

1.46 The characteristic polynomial of A is $(X-2)^4$, and the minimum polynomial is $(X-2)^2$. A has a single eigenvalue and is not diagonalisable. The possible Jordan normal forms are

$$\begin{bmatrix} 2 & 1 & 0 & 0 \\ 0 & 2 & 0 & 0 \\ 0 & 0 & 2 & 0 \\ 0 & 0 & 0 & 2 \end{bmatrix}, \quad \begin{bmatrix} 2 & 1 & 0 & 0 \\ 0 & 2 & 0 & 0 \\ 0 & 0 & 2 & 1 \\ 0 & 0 & 0 & 2 \end{bmatrix}.$$

Now $\dim \operatorname{Im}(A - 2I_4) = 1$ so $\dim \operatorname{Ker}(A - 2I_4) = 3$ and so the Jordan form is

$$\begin{bmatrix} 2 & 1 & 0 & 0 \\ 0 & 2 & 0 & 0 \\ 0 & 0 & 2 & 0 \\ 0 & 0 & 0 & 2 \end{bmatrix}.$$

Now $\text{Ker}(A - 2I_4) = \{[x, y, z, t] \mid 2x - y + t = 0\}$, and we must choose v_2 such that $(A - 2I_4)^2 v_2 = 0$ but $v_2 \notin \text{Ker}(A - 2I_4)$. So we take $v_2 = [1, 0, 0, 0]$, and then $v_1 = (A - 2I_4)v_2 = [-2, -2, -2, 2]$. We now wish to choose v_3 and v_4 such that $\{v_1, v_3, v_4\}$ is a basis for $\text{Ker}(A - 2I_4)$. So we take $v_3 = [0, 1, 0, 1]$ and $v_4 = [0, 0, 1, 0]$. Then we have

$$P = \begin{bmatrix} -2 & 1 & 0 & 0 \\ -2 & 0 & 1 & 0 \\ -2 & 0 & 0 & 1 \\ 2 & 0 & 1 & 0 \end{bmatrix}.$$

To solve the system $X' = AX$ we first solve the system $Y' = JY$, namely

$$y_1' = 2y_1 + y_2$$
$$y_2' = 2y_2$$
$$y_3' = 2y_3$$
$$y_4' = 2y_4.$$

The solution to this is clearly

$$y_4 = c_4 e^{2t}$$
$$y_3 = c_3 e^{2t}$$
$$y_2 = c_2 e^{2t}$$
$$y_1 = c_2 t e^{2t} + c_1 e^{2t}.$$

Since now

$$X = PY = \begin{bmatrix} -2 & 1 & 0 & 0 \\ -2 & 0 & 1 & 0 \\ -2 & 0 & 0 & 1 \\ 2 & 0 & 1 & 0 \end{bmatrix} \begin{bmatrix} c_2 t e^{2t} + c_1 e^{2t} \\ c_2 e^{2t} \\ c_3 e^{2t} \\ c_4 e^{2t} \end{bmatrix}$$

we deduce that

$$x_1 = -2c_2 t e^{2t} - 2c_1 e^{2t} + c_2 e^{2t}$$
$$x_2 = -2c_2 t e^{2t} - 2c_1 e^{2t} + c_3 e^{2t}$$
$$x_3 = -2c_2 t e^{2t} - 2c_1 e^{2t} + c_4 e^{2t}$$
$$x_4 = 2c_2 t e^{2t} + 2c_1 e^{2t} + c_3 e^{2t}.$$

Solutions to Chapter 1

1.47 (a) The system is $X' = AX$ where

$$A = \begin{bmatrix} 5 & 4 \\ -1 & 0 \end{bmatrix}.$$

The characteristic polynomial is $(X - 1)(X - 4)$. The eigenvalues are therefore 1 and 4, and associated eigenvectors are $E_1 = [1, -1]$ and $E_4 = [4, -1]$. The solution is $aE_1 e^t + bE_4 e^{4t}$, i.e.

$$x_1 = ae^t + 4be^{4t}$$
$$x_2 = -ae^t - be^{4t}.$$

(b) The system is $X' = AX$ where

$$A = \begin{bmatrix} 4 & -1 & -1 \\ 1 & 2 & -1 \\ 1 & -1 & 2 \end{bmatrix}.$$

The characteristic polynomial is $(X - 3)^2(X - 2)$. The eigenvalues are therefore 3 and 2. An eigenvector associated with 2 is $[1, 1, 1]$ so take $E_2 = [1, 1, 1]$. The eigenvalue 3 has geometric multiplicity 2 and $[1, 1, 0], [1, 0, 1]$ are linearly independent vectors in the eigenspace of 3. The general solution vector is therefore

$$a[1, 1, 1]e^{2t} + b[1, 1, 0]e^{3t} + c[1, 0, 1]e^{3t}$$

so that

$$x_1 = ae^{2t} + (b + c)e^{3t}$$
$$x_2 = ae^{2t} + be^{3t}$$
$$x_3 = ae^{2t} + ce^{3t}.$$

(c) The system is $X' = AX$ where

$$A = \begin{bmatrix} 5 & -6 & -6 \\ -1 & 4 & 2 \\ 3 & -6 & -4 \end{bmatrix}.$$

The characteristic polynomial is $(X - 1)(X - 2)^2$. The eigenvalues are therefore 1 and 2. An eigenvector associated with 1 is $E_1 = [3, -1, 3]$, and independent eigenvectors associated with 2 are $E_2 = [2, 1, 0]$ and $E_2' = [2, 0, 1]$. The solution space is then spanned by

$$\{[3e^t, -e^t, 3e^t], [2e^{2t}, e^{2t}, 0], [2e^{2t}, 0, e^{2t}]\}.$$

(d) The system is $X' = AX$ where

$$A = \begin{bmatrix} 1 & 3 & -2 \\ 0 & 7 & -4 \\ 0 & 9 & -5 \end{bmatrix}.$$

Now A has Jordan normal form

$$J = \begin{bmatrix} 1 & 1 & 0 \\ 0 & 1 & 0 \\ 0 & 0 & 1 \end{bmatrix}$$

and an invertible matrix P such that $P^{-1}AP = J$ is

$$P = \begin{bmatrix} 3 & 0 & 1 \\ 6 & 1 & 0 \\ 9 & 0 & 0 \end{bmatrix}.$$

First we solve $Y' = JY$ to obtain $y_1' = y_1 + y_2, y_2' = y_2, y_3' = y_3$ and hence

$$y_3 = ce^t$$
$$y_2 = be^t$$
$$y_1 = bte^t + ae^t.$$

Thus $X' = AX$ has the general solution

$$X = PY = a[3, 6, 9]e^t + b([3, 6, 9]te^t + [0, 1, 0]e^t) + c[1, 0, 0]e^t.$$

1.48 The system is $AY' = Y$ where

$$A = \begin{bmatrix} 1 & -3 & 2 \\ 0 & -5 & 4 \\ 0 & -9 & 7 \end{bmatrix}.$$

Now A is invertible with

$$A^{-1} = \begin{bmatrix} 1 & 3 & -2 \\ 0 & 7 & -4 \\ 0 & 9 & -5 \end{bmatrix}$$

and the system $Y' = A^{-1}Y$ is that of question 1.47(d).

Solutions to Chapter 1

1.49 The system is $X' = AX$ where

$$A = \begin{bmatrix} 1 & 1 \\ 2 & 3 \end{bmatrix}.$$

The eigenvalues are $2 + \sqrt{3}$ and $2 - \sqrt{3}$, with associated eigenvectors $[-1, -1 - \sqrt{3}]$ and $[-1, -1 + \sqrt{3}]$. The general solution is

$$a[-1, -1 - \sqrt{3}]e^{(2+\sqrt{3})t} + b[-1, -1 + \sqrt{3}]e^{(2-\sqrt{3})t}.$$

Since $x_1(0) = 0$ and $x_2(0) = 1$ we have $a + b = 0$ and $a - \sqrt{3}a - b + \sqrt{3}b = 1$, giving $a = -\dfrac{1}{2\sqrt{3}}$ and $b = \dfrac{1}{2\sqrt{3}}$, so the solution is

$$\frac{e^{2t}}{2\sqrt{3}}([1, 1 + \sqrt{3}]e^{\sqrt{3}t} + [-1, -1 + \sqrt{3}]e^{-\sqrt{3}t}).$$

1.50 Let $x = x_1, x_1' = x_2, x_1'' = x_2' = x_3, x_1''' = x_3' = 2x_3 + 4x_2 - 8x_1$. Then the system can be written in the form $X' = AX$ where

$$A = \begin{bmatrix} 0 & 1 & 0 \\ 0 & 0 & 1 \\ -8 & 4 & 2 \end{bmatrix}.$$

The characteristic polynomial is $(X - 2)(X^2 - 4)$ so the eigenvalues are 2 and -2. The Jordan normal form is

$$J = \begin{bmatrix} 2 & 1 & 0 \\ 0 & 2 & 0 \\ 0 & 0 & -2 \end{bmatrix}.$$

A Jordan basis $\{v_1, v_2, v_3\}$ satisfies

$$(A - 2I_3)v_1 = 0$$
$$(A - 2I_3)v_2 = v_1$$
$$(A + 2I_3)v_3 = 0.$$

Take $v_1 = [1, 2, 4]$ and $v_3 = [1, -2, 4]$. Then $v_2 = [0, 1, 4]$. Hence an invertible matrix P such that $P^{-1}AP = J$ is

$$P = \begin{bmatrix} 1 & 0 & 1 \\ 2 & 1 & -2 \\ 4 & 4 & 4 \end{bmatrix}.$$

Now solve the system $Y' = JY$ to get

$$y_1' = 2y_1 + y_2, y_2' = 2y_2, y_3' = -2y_3$$

so that $y_2 = c_2 e^{2t}, y_3 = c_3 e^{-2t}$ and hence $y_1' = 2y_1 + c_2 e^{2t}$ which gives $y_1 = c_1 e^{2t} + c_2 t e^{2t}$.

Now observe that

$$X = PY = \begin{bmatrix} 1 & 0 & 1 \\ 2 & 1 & -2 \\ 4 & 4 & 4 \end{bmatrix} \begin{bmatrix} c_1 e^{2t} + c_2 t e^{2t} \\ c_2 e^{2t} \\ c_3 e^{-2t} \end{bmatrix}.$$

Hence $x = x_1 = c_1 e^{2t} + c_2 t e^{2t} + c_3 e^{-2t}$. Now apply the initial conditions to obtain

$$x = (4t - 1)e^{2t} + e^{-2t}.$$

Solutions to Chapter 2

2.1   (a) $f \mapsto f'$ does not define a linear functional since $f' \notin \mathbb{R}$ in general.
(b),(c),(d) These are linear functionals.

(e) $\vartheta : f \mapsto \int_0^1 f^2$ is not a linear mapping; for example, we have $0 = \vartheta[f + (-f)]$ whereas in general

$$\vartheta(f) + \vartheta(-f) = 2 \int_0^1 f^2 \neq 0.$$

2.2   That φ is linear follows from the fact that

$$\varphi(\alpha f + \beta g) = \int_0^1 f_0(t)[\alpha f(t) + \beta g(t)]dt$$
$$= \alpha \int_0^1 f_0(t)\, f(t)\, dt + \beta \int_0^1 f_0(t)\, g(t)\, dt$$
$$= \alpha\varphi(f) + \beta\varphi(g).$$

2.3   The transition matrix from the given basis to the standard basis is

$$P = \begin{bmatrix} 1 & -1 & 0 \\ 0 & 1 & 1 \\ -1 & 0 & 1 \end{bmatrix}.$$

The inverse of this is readily seen to be

$$P^{-1} = \begin{bmatrix} \frac{1}{2} & \frac{1}{2} & -\frac{1}{2} \\ -\frac{1}{2} & \frac{1}{2} & -\frac{1}{2} \\ \frac{1}{2} & \frac{1}{2} & \frac{1}{2} \end{bmatrix}.$$

Hence the dual basis is

$$\{[\tfrac{1}{2}, \tfrac{1}{2}, -\tfrac{1}{2}], [-\tfrac{1}{2}, \tfrac{1}{2}, -\tfrac{1}{2}], [\tfrac{1}{2}, \tfrac{1}{2}, \tfrac{1}{2}]\}.$$

2.4   The transition matrix from the basis A' to the basis A is

$$P = \begin{bmatrix} 1 & 2 \\ 3 & 4 \end{bmatrix}.$$

Its inverse is

$$P^{-1} = \begin{bmatrix} -2 & 1 \\ \tfrac{3}{2} & -\tfrac{1}{2} \end{bmatrix}.$$

Consequently, $(A')^d = \{-2\varphi_1 + \tfrac{3}{2}\varphi_2, \varphi_1 - \tfrac{1}{2}\varphi_2\}.$

2.5   The transition matrix from the given basis to the standard basis is

$$P = \begin{bmatrix} -1 & 0 \\ 2 & 1 \end{bmatrix}.$$

Its inverse is

$$P^{-1} = \begin{bmatrix} -1 & 0 \\ 2 & 1 \end{bmatrix}$$

so the dual basis is (b), namely $\{[-1, 0], [2, 1]\}.$

2.6   (i) $\{[2, -1, 1, 0], [7, -3, 1, -1], [-10, 5, -2, 1], [-8, 3, -3, 1]\};$
       (ii) $\{(4, 5, -2, 11), (3, 4, -2, 6), (2, 3, -1, 4), (0, 0, 0, 1)\}.$

2.7   Since $V = A \oplus B$ we have

$$A^{\perp} + B^{\perp} = (A \cap B)^{\perp} = \{0\}^{\perp} = V^d$$
$$A^{\perp} \cap B^{\perp} = (A + B)^{\perp} = V^{\perp} = \{0\}.$$

Consequently, $V^d = A^{\perp} \oplus B^{\perp}.$

The answer to the question is 'no': A^d is the set of linear functionals $f : A \to F$ so if $A \neq V$ we have that A^d is not a subset of V^d. What is true is : if $V = A \oplus B$ then $V^d = A' \oplus B'$ where A', B' are subspaces of V^d with $A' \simeq A^d$ and $B' \simeq B^d$. To see this, let $f \in A^d$ and define $\overline{f} : V \to F$ by $\overline{f}(v) = f(a)$ where $v = a + b$. Then $\varphi : A^d \to V^d$ given by $\varphi(f) = \overline{f}$ is an injective linear transformation and $\varphi(A^d)$ is a subspace of V^d that is isomorphic to A^d. Define similarly $\mu : B^d \to V^d$ by $\mu(g) = \bar{g}$ where $\bar{g}(v) = g(b)$. Then we have

$$V^d = \varphi(A^d) \oplus \mu(B^d).$$

2.8 $\{f_1, f_2, f_3\}$ is linearly independent. For, if $\lambda_1 f_1 + \lambda_2 f_2 + \lambda_3 f_3 = 0$ then we have

$$0 = (\lambda_1 f_1 + \lambda_2 f_2 + \lambda_3 f_3)(1) = \lambda_1 + \lambda_2 + \lambda_3;$$
$$0 = (\lambda_1 f_1 + \lambda_2 f_2 + \lambda_3 f_3)(X) = \lambda_1 t_1 + \lambda_2 t_2 + \lambda_3 t_3;$$
$$0 = (\lambda_1 f_1 + \lambda_2 f_2 + \lambda_3 f_3)(X^2) = \lambda_1 t_1^2 + \lambda_2 t_2^2 + \lambda_3 t_3^2.$$

Since the coefficient matrix is the Vandermonde matrix and since the t_i are given to be distinct, the only solution is $\lambda_1 = \lambda_2 = \lambda_3 = 0$. Hence $\{f_1, f_2, f_3\}$ is linearly independent and so forms a basis for $(\mathbb{R}_3[X])^d$.

If $\{p_1, p_2, p_3\}$ is a basis of V of which $\{f_1, f_2, f_3\}$ is the dual then we must have $f_i(p_j) = \delta_{ij}$; i.e.

$$p_j(t_i) = \delta_{ij}.$$

It is now easily seen that

$$p_1(X) = \frac{(X - t_2)(X - t_3)}{(t_1 - t_2)(t_1 - t_3)}, \qquad p_2(X) = \frac{(X - t_1)(X - t_3)}{(t_2 - t_1)(t_2 - t_3)},$$
$$p_3(X) = \frac{(X - t_1)(X - t_2)}{(t_3 - t_1)(t_3 - t_2)}.$$

2.9   (a) $\alpha^{\wedge}(\varphi) = \varphi(\alpha) = \begin{bmatrix} 3 & 4 \end{bmatrix} \begin{bmatrix} 1 \\ 2 \end{bmatrix} = 11;$

     (b) $\beta^{\wedge}(\varphi) = \varphi(\beta) = \begin{bmatrix} 3 & 4 \end{bmatrix} \begin{bmatrix} 5 \\ 6 \end{bmatrix} = 39;$

     (c) $(2\alpha + 3\beta)^{\wedge}(\varphi) = \varphi(2\alpha + 3\beta) = \begin{bmatrix} 3 & 4 \end{bmatrix} \left(2 \begin{bmatrix} 1 \\ 2 \end{bmatrix} + 3 \begin{bmatrix} 5 \\ 6 \end{bmatrix} \right) = 139;$

     (d) $(2\alpha + 3\beta)^{\wedge} \left(\begin{bmatrix} a & b \end{bmatrix} \right) = \begin{bmatrix} a & b \end{bmatrix} \left(2 \begin{bmatrix} 1 \\ 2 \end{bmatrix} + 3 \begin{bmatrix} 5 \\ 6 \end{bmatrix} \right) = 17a + 22b.$

2.10   Let V be of dimension n and S of dimension k. Take a basis $\{v_1, \ldots, v_k\}$ of S and extend it to a basis

$$\{v_1, \ldots, v_k, v_{k+1}, \ldots, v_n\}$$

of V. Let $\{\varphi_1, \ldots, \varphi_n\}$ be the basis of V^d dual to $\{v_1, \ldots, v_n\}$. Given $\varphi \in V^d$ we have

$$\varphi = a_1 \varphi_1 + \cdots + a_n \varphi_n.$$

Since $\varphi(v_i) = a_i$ we see that $\varphi(v) = 0$ for every $v \in S$ if and only if $a_1 = \cdots = a_k = 0$. Thus

$$\varphi \in S^\perp \iff \varphi = a_{k+1}\varphi_{k+1} + \cdots + a_n\varphi_n$$

and so $\{\varphi_{k+1}, \ldots, \varphi_n\}$ is a basis for $S^\perp$. Hence $\dim S + \dim S^\perp = n$.

    If $\psi \in \operatorname{Ker} t^d$ then $t^d(\psi) = 0$ and so $[t^d(\psi)](u) = 0$ for all $u \in U$. But

$$[t^d(\psi)](u) = [\psi(t)](u) = \psi[t(u)].$$

Thus $\psi \in (\operatorname{Im} t)^\perp$. Conversely, let $\psi \in (\operatorname{Im} t)^\perp$. Then for all $u \in U$ we have

$$[t^d(\psi)](u) = \psi[t(u)] = 0$$

and so $t^d(\psi) = 0$ whence $\psi \in \operatorname{Ker} t^d$. Thus $\operatorname{Ker} t^d = (\operatorname{Im} t)^\perp$.

    (i) means that $v \in \operatorname{Im} t$ while (ii) means $v \notin (\operatorname{Ker} t^d)^\perp$. In terms of linear equations, this says that either (i) the system $AX = B$ has a solution, or (ii) there is a row vector C with $CA = 0$ and $CB = 1$.

    It is readily seen that the given system of equations has no solution, so (ii) holds. The linear functional satisfying (ii) is $[1, -2, 1]$.

2.11   If $\varphi \in W^d$ then for all $u \in U$ we have

$$(s \circ t)^d(\varphi)(u) = \varphi[(s \circ t)(u)] = \varphi s[t(u)] = s^d(\varphi)[t(u)] = [t^d(s^d(\varphi))](u)$$

from which the result follows. The final statements are immediate from the fact that $\operatorname{Im} t = (\operatorname{Ker} t^d)^\perp$ (see question 2.10).

2.12   To find Y we find the dual of $\{[1, 0, 0], [1, 1, 0], [1, 1, 1]\}$. The transition matrix and its inverse are

$$P = \begin{bmatrix} 1 & 1 & 1 \\ 0 & 1 & 1 \\ 0 & 0 & 1 \end{bmatrix}, \qquad P^{-1} = \begin{bmatrix} 1 & -1 & 0 \\ 0 & 1 & -1 \\ 0 & 0 & 1 \end{bmatrix}.$$

Thus $Y = \{(1, -1, 0), (0, 1, -1), (0, 0, 1)\}$.

The matrix of t with respect to the standard basis is

$$\begin{bmatrix} 2 & 1 & 0 \\ 1 & 1 & 1 \\ 0 & 0 & -1 \end{bmatrix}$$

and the transition matrices relative to X, Y are respectively

$$\begin{bmatrix} 1 & 0 & 0 \\ -1 & 1 & 0 \\ 0 & -1 & 1 \end{bmatrix}, \quad \begin{bmatrix} 1 & 1 & 1 \\ 0 & 1 & 1 \\ 0 & 0 & 1 \end{bmatrix}.$$

By the change of basis theorem, the matrix of t relative to X, Y is then

$$\begin{bmatrix} 1 & 0 & 0 \\ 1 & 1 & 0 \\ 1 & 1 & 1 \end{bmatrix}\begin{bmatrix} 2 & 1 & 0 \\ 1 & 1 & 1 \\ 0 & 0 & -1 \end{bmatrix}\begin{bmatrix} 1 & 1 & 1 \\ 0 & 1 & 1 \\ 0 & 0 & 1 \end{bmatrix} = \begin{bmatrix} 2 & 3 & 3 \\ 3 & 5 & 6 \\ 3 & 5 & 5 \end{bmatrix}.$$

The required matrix is then the transpose of this one.

2.13 The annihilator $\langle \alpha_1, \alpha_2 \rangle^\perp$ is of dimension 1 and contains both φ_3 and φ'_3. Thus φ'_3 is a scalar multiple of φ_3.

2.14 It is immediate from properties of integrals that

$$L_g(\lambda_1 f_1 + \lambda_2 f_2) = \lambda_1 L_g(f_1) + \lambda_2 L_g(f_2)$$

and so L_g is linear.

To show that F_x is a linear functional, we must check that

$$F_x(\lambda_1 f_1 + \lambda_2 f_2) = \lambda_1 F_x(f_1) + \lambda_2 F_x(f_2),$$

i.e. that $(\lambda_1 f_1 + \lambda_2 f_2)(x) = \lambda_1 f_1(x) + \lambda_2 f_2(x)$, which is clear.

Suppose that x is fixed and that $F_x = L_g$ for some g. Then

$$(\forall f \in C[0,1]) \qquad \int_0^1 f(t)\, g(t)\, dt = f(x).$$

Now, by continuity, the left hand side depends on the values of f at points other than x whereas the right hand side does not. Thus $F_x \neq L_g$ for any g.

2.15   The method of solution is outlined in the question.

2.16   Let $\{v_1, \ldots, v_{k-1}\}$ be a basis of U and extend this to a basis $\{v_1, \ldots, v_k\}$ of V. Apply the Gram–Schmidt process to obtain an orthonormal basis $\{u_1, \ldots, u_k\}$ of V, where $\{u_1, \ldots, u_{k-1}\}$ is an orthonormal basis of U. Let $n = u_k$. Then if $x = \sum_{i=1}^{k-1} \lambda_i u_i \in U$ we have

$$\langle n|x \rangle = \langle u_k | \lambda_1 u_1 + \cdots + \lambda_{k-1} u_{k-1} \rangle = \sum_{i=1}^{k-1} \lambda_i \langle u_k | u_i \rangle = 0.$$

Conversely, if $x = \sum_{i=1}^{k} \lambda_i u_i \in V$ and if $\langle n|x \rangle = 0$ then

$$0 = \langle u_k | \lambda_1 u_1 + \cdots + \lambda_k u_k \rangle = \sum_{i=1}^{k} \lambda_i \langle u_k | u_i \rangle = \lambda_k.$$

Consequently we see that $x \in U$ and so

$$U = \{x \in V \mid \langle n|x \rangle = 0\}.$$

Now $v - v' = 2\langle n|v \rangle n$, a scalar multiple of n, and so is orthogonal to U. Then $\frac{1}{2}(v + v') = v - \langle n|v \rangle n \in U$ since

$$\langle n|v - \langle n|v \rangle n \rangle = \langle n|v \rangle - \langle n|v \rangle \langle n|n \rangle$$
$$= \langle n|v \rangle - \langle n|v \rangle$$
$$= 0.$$

We have

$$t(v + w) = v + w - 2\langle n|v + w \rangle n$$
$$= v + w - 2(\langle n|v \rangle + \langle n|w \rangle)n$$
$$= (v - 2\langle n|v \rangle n) + (w - 2\langle n|v \rangle n)$$
$$= t(v) + t(w);$$
$$t(\lambda v) = \lambda v - 2\langle n|\lambda v \rangle n$$
$$= \lambda v - 2\lambda \langle n|v \rangle n$$
$$= \lambda t(v),$$

and so t is linear.

Note that $t(u) = u$ for every $u \in U$ and so 1 is an eigenvalue and $\mathrm{Ker}(t - \mathrm{id}) = U$ is of dimension $k - 1$. Thus 1 has geometric multiplicity $k - 1$. Also, $t(n) = -n$ and so -1 is also an eigenvalue with associated eigenvector n. Since the sum of the geometric multiplicities is k it follows that t is diagonalisable.

In the last part we have that $n = \frac{1}{\sqrt{11}}(3, -1, 1)$ so if $v = (a, b, c)$ then $\langle n, v \rangle = \frac{1}{\sqrt{11}}(3a - b + c)$. Thus

$$s(a, b, c) = (a, b, c) - \frac{2}{11}(3a - b + c)(3, -1, 1)$$
$$= \frac{1}{11}(-7a + 6b - 6c, 6a + 9b + 2c, -6a + 2b + 9c),$$

which gives

$$\text{mat } s = \frac{1}{11}\begin{bmatrix} -7 & 6 & -6 \\ 6 & 9 & 2 \\ -6 & 2 & 9 \end{bmatrix}.$$

As for t, we have $n = \frac{1}{\sqrt{10}}(2, -1, 2, -1)$ and so if $v = (a, b, c, d)$ then

$$t(a, b, c, d) = (a, b, c, d) - \frac{1}{5}(2a - b + 2c - d)(2, -1, 2, -1)$$
$$= \frac{1}{5}(a + 2b - 4c + 2d, 2a + 4b + 2c - d,$$
$$- 4a + 2b + c + 2d, 2a - b + 2c + 4d),$$

which gives

$$\text{mat } t = \frac{1}{5}\begin{bmatrix} 1 & 2 & -4 & 2 \\ 2 & 4 & 2 & -1 \\ -4 & 2 & 1 & 2 \\ 2 & -1 & 2 & 4 \end{bmatrix}.$$

2.17 The hyperbola is represented by the equation

$$\begin{bmatrix} x & y \end{bmatrix}\begin{bmatrix} -1 & 3 \\ 3 & -1 \end{bmatrix}\begin{bmatrix} x \\ y \end{bmatrix} = 1.$$

The eigenvalues of $A = \begin{bmatrix} -1 & 3 \\ 3 & -1 \end{bmatrix}$ are 2 and -4 with associated normalised eigenvectors

$$\begin{bmatrix} 1/\sqrt{2} \\ 1/\sqrt{2} \end{bmatrix}, \qquad \begin{bmatrix} -1/\sqrt{2} \\ 1/\sqrt{2} \end{bmatrix}.$$

Thus $P^t A P = \text{diag}\{2, -4\}$ where

$$P = \begin{bmatrix} 1/\sqrt{2} & -1/\sqrt{2} \\ 1/\sqrt{2} & 1/\sqrt{2} \end{bmatrix}.$$

Now change coordinates by defining

$$\mathbf{x}_1 = \begin{bmatrix} x_1 \\ y_1 \end{bmatrix} = P^t \begin{bmatrix} x \\ y \end{bmatrix} = P^t \mathbf{x}.$$

Then we have $\mathbf{x} = P\mathbf{x}_1$ and the original equation becomes

$$1 = \mathbf{x}^t A \mathbf{x} = (P\mathbf{x}_1)^t A (P\mathbf{x}_1)$$
$$= \mathbf{x}_1^t P^t A P \mathbf{x}_1$$
$$= \begin{bmatrix} x_1 & y_1 \end{bmatrix} \begin{bmatrix} 2 & 0 \\ 0 & -4 \end{bmatrix} \begin{bmatrix} x_1 \\ y_1 \end{bmatrix}$$
$$= 2x_1^2 - 4y_1^2.$$

Thus the principal axes are given by $x_1 = 0$ and $y_1 = 0$, i.e. $y = -x$ and $y = x$.

The ellipsoid is represented by the equation

$$\begin{bmatrix} x & y & z \end{bmatrix} \begin{bmatrix} 7 & 2 & 0 \\ 2 & 6 & -2 \\ 0 & -2 & 5 \end{bmatrix} \begin{bmatrix} x \\ y \\ z \end{bmatrix} = 1.$$

The eigenvalues of $A = \begin{bmatrix} 7 & 2 & 0 \\ 2 & 6 & -2 \\ 0 & -2 & 5 \end{bmatrix}$ are $3, 6, 9$ with associated normalised eigenvectors

$$\begin{bmatrix} -1/3 \\ 2/3 \\ 2/3 \end{bmatrix}, \qquad \begin{bmatrix} 2/3 \\ -1/3 \\ 2/3 \end{bmatrix}, \qquad \begin{bmatrix} 2/3 \\ 2/3 \\ -1/3 \end{bmatrix}.$$

Thus $P^t A P = \operatorname{diag}\{3, 6, 9\}$ where

$$P = \frac{1}{3} \begin{bmatrix} -1 & 2 & 2 \\ 2 & -1 & 2 \\ 2 & 2 & -1 \end{bmatrix}.$$

Now change coordinates by defining

$$\mathbf{x}_1 = \begin{bmatrix} x_1 \\ y_1 \\ z_1 \end{bmatrix} = P^t \begin{bmatrix} x \\ y \\ z \end{bmatrix} = P^t \mathbf{x}.$$

Then we have $\mathbf{x} = P\mathbf{x}_1$ and the original equation becomes (with a calculation similar to the above)

$$3x_1^2 + 6y_1^2 + 9z_1^2 = 1.$$

Since

$$x_1 = \tfrac{1}{3}(-x + 2y + 2z),$$
$$y_1 = \tfrac{1}{3}(2x - y + 2z),$$
$$z_1 = \tfrac{1}{3}(2x + 2y - z),$$

the x_1-axis is given by $y_1 = z_1 = 0$ and has direction numbers $(-1, 2, 2)$;
the y_1-axis is given by $x_1 = z_1 = 0$ and has direction numbers $(2, -1, 2)$;
the z_1-axis is given by $x_1 = y_1 = 0$ and has direction numbers $(2, 2, -1)$.

2.18 Let $\{v_1, \ldots, v_n\}$ be an orthonormal basis of V. Then for every $x \in V$ we have $x = \sum_{k=1}^{n} \langle x|v_k \rangle v_k$ so in particular

$$(i = 1, \ldots, n) \qquad f(v_i) = \sum_{k=1}^{n} \langle f(v_i)|v_k \rangle v_k.$$

If A is the matrix of f we thus see that $a_{jk} = \langle f(v_j)|v_k \rangle$. If M is the matrix of f^* then likewise we have that $m_{jk} = \langle f^*(v_j)|v_k \rangle$. The result now follows from the observation that

$$m_{jk} = \langle f^*(v_j)|v_k \rangle = \overline{\langle v_k|f^*(v_j) \rangle}$$
$$= \overline{\langle f(v_k)|v_j \rangle}$$
$$= \bar{a}_{kj}.$$

2.19 The first part is a routine check of the axioms. Using the fact that $\operatorname{tr}(AB) = \operatorname{tr}(BA)$ we have

$$\langle f_M(A)|B \rangle = \operatorname{tr}[B^*(MA)] = \operatorname{tr}[MAB^*]$$
$$= \operatorname{tr}[B^*MA]$$
$$= \operatorname{tr}[(M^*B)^*A]$$
$$= \langle A|f_{M^*}(B) \rangle$$

from which it follows that $(f_M)^* = f_{M^*}$.

2.20 Let β be as stated and let $f_\beta \in V^\star$ be given by $f_\beta(\alpha) = \langle \alpha | \beta \rangle$. Then since $\{\alpha_1, \ldots, \alpha_n\}$ is an orthonormal basis we have

$$f_\beta(\alpha_k) = \langle \alpha_k | \sum_{i=1}^n \overline{f(\alpha_i)} \alpha_i \rangle = f(\alpha_k).$$

Since f_β and f coincide on this basis, it follows that $f = f_\beta$.

For the next part suppose that such a q exists. Then we have

$$p(z) = f(p) = \int_0^1 p(t)\,\overline{q(t)}\,dt$$

for every p. Writing rp for p we then have

$$\int_0^1 r(t)\,p(t)\,\overline{q(t)}\,dt = r(z)p(z) = 0.$$

In particular, this holds when $p = \overline{r}q$. So

$$\int_0^1 |r(t)|^2\,|q(t)|^2\,dt = 0$$

whence we have $rq = 0$. Since $r \neq 0$ we must therefore have $q = 0$, a contradiction.

For the next part of the question note that

$$\begin{aligned}
\langle f_p(q) | r \rangle &= \langle pq | r \rangle \\
&= \int_0^1 p(t)\,q(t)\,\overline{r(t)}\,dt \\
&= \int_0^1 q(t)\,\overline{\overline{p(t)}\,r(t)}\,dt \\
&= \langle q | \overline{p}r \rangle \\
&= \langle q | f_{\overline{p}}(r) \rangle
\end{aligned}$$

so $(f_p)^\star = f_{\overline{p}}$.

Integration by parts gives

$$\langle D(p) | q \rangle = p(1)\overline{q}(1) - p(0)\overline{q}(0) - \langle p | D(q) \rangle.$$

If $D^\star$ exists then we have

$$\langle p | D^\star(q) \rangle = p(1)\overline{q}(1) - p(0)\overline{q}(0) - \langle p | D(q) \rangle$$

so that

$$\langle p | D(q) + D^\star(q) \rangle = p(1)\overline{q}(1) - p(0)\overline{q}(0).$$

Now fix q such that $q(0) = 0, q(1) = 1$. Then we have

$$\langle p | D(q) + D^\star(q) \rangle = p(1),$$

which is impossible (take $z = 1$ in the previous part of the question). Thus $D^\star$ does not exist.

2.21 That K is self-adjoint follows from the fact that xy is symmetric in x and y.

We have

$$
\begin{aligned}
K(f_n) &= \int_0^1 xy\, f_n(y)\, dy \\
&= \int_0^1 xy\left(y^n - \frac{2}{n+2}\right) dy \\
&= \int_0^1 xy^{n+1}\, dy - \frac{2}{n+2}\int_0^1 xy\, dy \\
&= \frac{xy^{n+2}}{n+2}\bigg|_{y=0}^{y=1} - \frac{2}{n+2}\frac{xy^2}{2}\bigg|_{y=0}^{y=1} \\
&= 0,
\end{aligned}
$$

so $K(f_n) = 0 f_n$ as required.

Apply the Gram–Schmidt process to $\{f_1, f_2\}$. Let $e_1(x) = f_1(x) = x - \frac{2}{3}$ and define $e_2(x) = f_2(x) + \alpha f_1(x) = (x^2 - \frac{1}{2}) + \alpha(x - \frac{2}{3})$ where

$$
\alpha = -\frac{\langle\, x^2 - \frac{1}{2}, x - \frac{2}{3}\,\rangle}{\langle\, x - \frac{2}{3}, x - \frac{2}{3}\,\rangle}.
$$

Since

$$
\langle\, x^2 - \tfrac{1}{2}, x - \tfrac{2}{3}\,\rangle = \int_0^1 (x^2 - \tfrac{1}{2})(x - \tfrac{2}{3})\, dx = \tfrac{1}{9},
$$

$$
\langle\, x - \tfrac{2}{3}, x - \tfrac{2}{3}\,\rangle = \int_0^1 (x - \tfrac{2}{3})^2\, dx = \tfrac{1}{9}
$$

it follows that $\alpha = -1$ and hence that

$$
e_2(x) = x^2 - x + \tfrac{1}{6}.
$$

Thus e_1, e_2 are orthogonal eigenfunctions associated with the eigenvalue 0.

If $K(f) = \lambda f$ then

$$
\lambda f(x) = x \int_0^1 y f(y)\, dy.
$$

If $\lambda \neq 0$ then f must then be of the form $x \mapsto \alpha x$ for some constant α. Substituting αx for $f(x)$, it clearly follows that $\lambda = \frac{1}{3}$. An associated eigenfunction is given by $f(x) = x$.

2.22 Since t is skew-adjoint its eigenvalues are purely imaginary so ± 1 are not eigenvalues. Hence $\mathrm{id} \pm t$ cannot be singular. Now

$$s^* = (\mathrm{id} + t^*)^{-1}(\mathrm{id} - t^*) = (\mathrm{id} - t)^{-1}(\mathrm{id} + t).$$

But $(\mathrm{id} + t)(\mathrm{id} - t) = \mathrm{id} - t^2 = (\mathrm{id} - t)(\mathrm{id} + t)$ so

$$(\mathrm{id} - t)^{-1}(\mathrm{id} + t) = (\mathrm{id} + t)(\mathrm{id} - t)^{-1}.$$

Hence $s^* = (\mathrm{id} + t)(\mathrm{id} - t)^{-1}$ and so

$$ss^* = (\mathrm{id} - t)(\mathrm{id} + t)^{-1}(\mathrm{id} + t)(\mathrm{id} - t)^{-1} = \mathrm{id}.$$

To see that s cannot have -1 as an eigenvalue, consider

$$\mathrm{id} + s = \mathrm{id} + (\mathrm{id} - t)(\mathrm{id} + t)^{-1}.$$

We have that

$$(\mathrm{id} + s)(\mathrm{id} + t) = (\mathrm{id} + t) + (\mathrm{id} - t) = 2\,\mathrm{id},$$

and so $(\mathrm{id} + s)^{-1} = \frac{1}{2}(\mathrm{id} + t)$ whence $\mathrm{id} + s$ does not have 0 as an eigenvalue and hence -1 is not an eigenvalue of s.

2.23 $S^t = S$ and $T^t = -T$, so

$$(\overline{T + iS})^t = (T - iS)^t = T^t - iS^t = -(T + iS).$$

Thus $T + iS$ is skew-adjoint. But the eigenvalues of a skew-adjoint matrix are purely imaginary, so 1 is not an eigenvalue of $T + iS$, so

$$\det(T + iS - I) \neq 0.$$

As for the second part, we have

$$\begin{aligned}
U &= (I + T + iS)(I - T - iS)^{-1}\\
&= [I + (T + iS)][I - (T + iS)]^{-1}.
\end{aligned}$$

The fact that U is unitary now follows from the previous question.

2.24 It is given that $A^t = A, S^t = -S, AS = SA, \det(A - S) \neq 0$. Let $B = (A + S)(A - S)^{-1}$. Then we have

$$
\begin{aligned}
B^t B &= [(A + S)(A - S)^{-1}]^t (A + S)(A - S)^{-1} \\
&= [(A - S)^{-1}]^t (A + S)^t (A + S)(A - S)^{-1} \\
&= (A^t - S^t)^{-1}(A^t + S^t)(A + S)(A - S)^{-1} \\
&= (A + S)^{-1}(A - S)(A + S)(A - S)^{-1} \\
&= (A + S)^{-1}(A + S)(A - S)(A - S)^{-1},
\end{aligned}
$$

the last equality following from the fact that since A, S commute so do $A + S$ and $A - S$. Hence $B^t B = I$ and B is orthogonal.

2.25 Since

(1)
$$\overline{A}^t A + A = 0$$

we have, taking transposes,

$$A^t \overline{A} + A^t = 0$$

and hence, taking complex conjugates,

(2)
$$\overline{A}^t A + \overline{A}^t = 0.$$

It follows from (1) and (2) that $\overline{A}^t = A$, so that A is self-adjoint. Let $\lambda_1, \ldots, \lambda_n$ be the distinct non-zero eigenvalues of A. Then $\lambda_1, \ldots, \lambda_n$ are necessarily real.

The relation (1) can now be written in the form

$$A^2 = -A$$

from which it follows that the distinct non-zero eigenvalues of $-A$, namely $-\lambda_1, \ldots, -\lambda_n$ are precisely the distinct non-zero eigenvalues of A^2, namely $\lambda_1^2, \ldots, \lambda_n^2$. It follows that $\lambda_1, \ldots, \lambda_n$ are all negative. Let $\alpha_i = -\lambda_i$ for each i, and suppose that

$$\alpha_1 < \alpha_2 < \cdots < \alpha_n.$$

Then this chain must coincide with the chain

$$\alpha_1^2 < \alpha_2^2 < \cdots < \alpha_n^2.$$

Consequently, $\alpha_i = \alpha_i^2$ for every i and, since by hypothesis $\alpha_i \neq 0$, we obtain $\lambda_i = -\alpha_i = -1$.

2.26 A and B are given to be orthogonal with $\det A + \det B = 0$. Now we have that

(1) $$\det(A + B) = \det[(AB^t + I)B]$$

from which we see that $\det(A + B) = 0$ if and only if -1 is an eigenvalue of AB^t. But AB^t is orthogonal since

$$(AB^t)^t(AB^t) = BA^t AB^t = I.$$

Also, $\det(AB^t) = \det A \det B^{-1} = -1$ since it is given that $\det A = -\det B$. Thus $C = AB^t$ is orthogonal with $\det C = -1$. It follows that -1 is an eigenvalue of C; for

$$C^t(I + C) = C^t + I = (C + I)^t$$

and so, taking determinants,

$$\det(I + C)[\det C - 1] = 0$$

whence $\det(I + C) = 0$. It now follows from (1) that $\det(A + B) = 0$.

2.27 (1) Since $A^t(A - I) = I - A^t = -(A^t - I)$ we have that

$$\det A \det(A - I) = (-1)^n \det(A - I)$$

and so

$$\det(A - I)[\det A - (-1)^n] = 0.$$

If $\det A = 1$ and n is odd then it follows that $\det(A - I) = 0$ and hence 1 is an eigenvalue of A. If $\det A = -1$ and n is even then likewise $\det(A - I) = 0$ and again 1 is an eigenvalue of A.

(2) Since $A^t(I + A) = A^t + I = (I + A)^t$ we have that

$$\det A \det(I + A) = \det(I + A)$$

and so if $\det A = -1$ then $\det(I + A) = 0$ whence -1 is an eigenvalue of A.

2.28 The first part follows from the observation that

$$g(A) = 0 \iff g(A^t) = 0 \iff g(-A) = 0.$$

Suppose now that $g(X)$ is the minimum polynomial of A, say

$$g(X) = a_0 + a_1 X + \cdots + a_{r-1} X^{r-1} + X^r.$$

Since $g(-A) = 0$ we have that

$$a_0 - a_1 X + \cdots + (-1)^r X^r$$

is also the minimum polynomial of A. Thus $a_1 = a_3 = \cdots = 0$.

Since $(A^n)^t = (-1)^n A$ for a skew-symmetric matrix A we see that A^n is skew-symmetric if n is odd, and is symmetric if n is even. Hence $f(A)$ is skew-symmetric, and $g(A)$ is symmetric.

2.29 We have

$$N(UA) = \text{tr}(\overline{UA}^t UA)$$
$$= \text{tr}(\overline{A}^t \overline{U}^t UA)$$
$$= \text{tr}(\overline{A}^t A)$$
$$= N(A).$$

Similarly, using the fact that $\text{tr}(XY) = \text{tr}(YX)$,

$$N(AU) = \text{tr}[(\overline{AU})^t AU]$$
$$= \text{tr}[AU(\overline{AU})^t]$$
$$= \text{tr}(AU\overline{U}^t \overline{A}^t)$$
$$= \text{tr}(A\overline{A}^t)$$
$$= N(A).$$

Finally, by the above,

$$N(I_n - U^{-1}A) = N[U(I_n - U^{-1}A)]$$
$$= N(U - A)$$
$$= N(A - U).$$

2.30 Since $\overline{A}^t A = A\overline{A}^t$ we have that $A^{-1}(\overline{A}^t)^{-1} = (\overline{A}^t)^{-1}A^{-1}$. But

$$A^{-1}A = I \Longrightarrow \overline{A^{-1}}\,\overline{A} = I \Longrightarrow \overline{A}^{-1} = \overline{A^{-1}}.$$

It follows that

$$A^{-1}\overline{A^{-1}}^t = A^{-1}(\overline{A}^t)^{-1} = (\overline{A}^t)^{-1}A^{-1} = \overline{A^{-1}}^t A^{-1}$$

and so A^{-1} is normal.

If $\overline{A}^t = a_0 I + a_1 A + \cdots + a_n A^n$ then clearly $A\overline{A}^t = \overline{A}^t A$. Suppose conversely that A is normal. Then there is a unitary matrix P and a diagonal matrix D such that

$$A = P^{-1}DP = \overline{P}^t DP, \qquad \overline{A}^t = \overline{P}^t \overline{D}P = P^{-1}\overline{D}P.$$

Let $\lambda_1, \ldots, \lambda_r$ be the distinct elements of D. Consider the equations

$$\overline{\lambda}_1 = a_0 + a_1\lambda_1 + a_2\lambda_1^2 + \cdots + a_{r-1}\lambda_1^{r-1}$$
$$\overline{\lambda}_2 = a_0 + a_1\lambda_2 + a_2\lambda_2^2 + \cdots + a_{r-1}\lambda_2^{r-1}$$
$$\vdots$$
$$\overline{\lambda}_r = a_0 + a_1\lambda_r + a_2\lambda_r^2 + \cdots + a_{r-1}\lambda_r^{r-1}.$$

Since $\lambda_1, \ldots, \lambda_r$ are distinct the (Vandermonde) coefficient matrix has non-zero determinant and so the system has a unique solution. We then have

$$\overline{D} = a_0 I + a_1 D + a_2 D^2 + \cdots + a_{r-1} D^{r-1}$$

and consequently

$$\overline{A}^t = P^{-1} \overline{D} P = P^{-1}(a_0 I + a_1 D + \cdots + a_{r-1} D^{r-1})P$$
$$= a_0 I + a_1 A + \cdots + a_{r-1} A^{r-1}.$$

2.31 Suppose that A is normal and let $B = g(A)$. There is a unitary matrix P and a diagonal matrix D such that

$$A = \overline{P}^t DP = P^{-1} DP.$$

Consequently we have

$$B = g(A) = \overline{P}^t g(D)P = P^{-1} g(D)P,$$

and so

$$\overline{B}^t B = \overline{P}^t g(\overline{D})P \overline{P}^t g(D)P = \overline{P}^t g(\overline{D})g(D)P$$

and similarly

$$B \overline{B}^t = \overline{P}^t g(D) g(\overline{D})P.$$

Since $g(D)$ and $g(\overline{D})$ are diagonal matrices, it follows that $\overline{B}^t B = B \overline{B}^t$ and so B is normal.

2.32 We have that

$$(A + Bi)^\star (A + Bi) = (A^\star - B^\star i)(A + Bi)$$
$$= (A - Bi)(A + Bi)$$
$$= A^2 - (BA - AB)i + B^2,$$

and similarly $(A + Bi)(A + Bi)^\star = A^2 - (AB - BA)i + B^2$. It follows that $A + Bi$ is normal if and only if $AB = BA$.

2.33 To get $-A$ multiply each row of A by -1. Then clearly $\det(-A) = (-1)^n \det A$. If n is odd then

$$\det A = \det(A^t) = \det(-A) = (-1)^n \det A = -\det A$$

and so $\det A = 0$.

Since $x^t A x$ is a 1×1 matrix we have

$$x^t A x = (x^t A x)^t = x^t A^t x = -x^t A x$$

and so $x^t A x = 0$.

Let $A x = \lambda x$ and let stars denote transposes of complex conjugates. Then we have $x^* A x = \lambda x^* x$. Taking the star of each side and using $A^* = A^t = -A$, we obtain

$$\overline{\lambda} x^* x = (x^* A x)^* = x^* A^* x = -x^* A x = -\lambda x^* x.$$

Since $x^* x \neq 0$ it follows that $\overline{\lambda} = -\lambda$. Thus $\lambda = i\mu$ where $\mu \in \mathbb{R} \setminus \{0\}$.

If $x = y + iz$ then from $A x = i\mu x$ we obtain $A(y + iz) = i\mu(y + iz)$ and so, equating real and imaginary parts, $Ay = -\mu z$, $Az = \mu y$. Now

$$\mu y^t y = y^t A z = (y^t A z)^t = z^t A^t y = -z^t A y = \mu z^t z$$

and so $y^t y = z^t z$. Also, $\mu y^t z = -y^t A y = 0$ (by the first part of the question). If, therefore, $Au = 0$ then

$$\mu u^t y = u^t A z = -(Au)^t z = 0$$

and similarly

$$\mu u^t z = -u^t A y = (Au)^t y = 0.$$

For the last part, we have

$$\det(A - \lambda I) = \det \begin{bmatrix} -\lambda & 2 & -2 \\ -2 & -\lambda & -1 \\ 2 & 1 & -\lambda \end{bmatrix} = -\lambda(\lambda^2 + 9),$$

so the eigenvalues are 0 and $\pm 3i$.

A normalised eigenvector corresponding to 0 is

$$u = \frac{1}{3} \begin{bmatrix} -1 \\ 2 \\ 2 \end{bmatrix}.$$

To find y, z as above, choose y perpendicular to u, say

$$y = \frac{1}{\sqrt{2}} \begin{bmatrix} 0 \\ -1 \\ 1 \end{bmatrix}.$$

Then we have

$$-3z = Ay = \frac{1}{\sqrt{2}} \begin{bmatrix} -4 \\ -1 \\ -1 \end{bmatrix}$$

which gives

$$z = \frac{1}{3\sqrt{2}} \begin{bmatrix} 4 \\ 1 \\ 1 \end{bmatrix}.$$

Relative to the basis $\{u, y, z\}$ the representing matrix is now

$$\begin{bmatrix} 0 & 0 & 0 \\ 0 & 0 & 3 \\ 0 & -3 & 0 \end{bmatrix}.$$

The required orthogonal matrix P is then

$$P = \begin{bmatrix} -1/3 & 0 & 4/3\sqrt{2} \\ 2/3 & -1/\sqrt{2} & 1/3\sqrt{2} \\ 2/3 & 1/\sqrt{2} & 1/3\sqrt{2} \end{bmatrix}.$$

2.34 Let Q be the matrix that represents the change to a new basis with respect to which q is in normal form. Then $x^t A x$ becomes $y^t B y$ where $x = Qy$ and $B = Q^t A Q$. Now

$$q(x) = q'(y) = y_1^2 + \cdots + y_p^2 - y_{p+1}^2 - \cdots - y_{p+m}^2$$

where $p - m$ is the signature of q and $p + m$ is the rank of q. Notice that the rank of q is equal to the rank of the matrix B which is in turn equal to the rank of the matrix A (since Q is non-singular), and

$$y = (y_1, \ldots, y_p, y_{p+1}, \ldots, y_{p+m}, y_{p+m+1}, \ldots, y_n)^t.$$

Now if q has the same rank and signature then clearly $m = 0$. Hence $y^t B y \geq 0$ for all $y \in \mathbb{R}^n$ since it is a sum of squares. Consequently $x^t A x \geq 0$ for all $x \in \mathbb{R}^n$.

Conversely, if $x^t A x \geq 0$ for all $x \in \mathbb{R}^n$ then $y^t B y \geq 0$ for all $y \in \mathbb{R}^n$. Choose $y = (0, \ldots, 0, y_i, 0, \ldots, 0)$. Now the coefficient of y_i^2 must be 0 or 1, but not -1. Therefore there are no terms of the form $-y_i^2$, so $m = 0$ and q has the same rank and signature.

If the rank and signature are both equal to n then $m = 0$ and $p = n$. Hence

$$y^t B y = y_1^2 + \cdots + y_n^2.$$

But a sum of squares is zero if and only if each term is zero, so $x^t A x \geq 0$ and is equal to 0 only when $x = 0$.

Conversely, if $x^t A x \geq 0$ for $x \in \mathbb{R}^n$ then $y^t B y \geq 0$ for $y \in \mathbb{R}^n$ so $m = 0$, for otherwise we can choose

$$y = (0, \ldots, 0, y_{p+1}, 0, \ldots, 0)$$

with $y_{p+1} = 1$ to obtain $y^t B y < 0$. Also, $x^t A x = 0$ only for $x = 0$ gives $y^t B y = 0$ only for $y = 0$. If $p < n$ then, since we have $m = 0$, choose $y = (0, \ldots, 0, 1)$ to get $y^t B y = 0$ with $y \neq 0$. Hence $p = n$ as required.

2.35 The quadratic form q can be reduced to normal form either by completing squares or by row and column operations. We solve the problem by completing squares. We have

$$q(x) = x_1^2 + 2x_1 x_2 + x_2^2 - 2x_1 x_3 - x_3^2$$
$$= (x_1 + x_2)^2 + x_1^2 - (x_1 + x_3)^2$$

and so the normal form of q is

$$\begin{bmatrix} 1 & 0 & 0 \\ 0 & 1 & 0 \\ 0 & 0 & -1 \end{bmatrix}.$$

Since the rank of q is 3 and its signature is 1, q is neither positive definite nor positive semi-definite.

Coordinates (x_1, x_2, x_3) with respect to the standard basis become $(x_1 + x_2, x_1, x_1 + x_3)$ in the new basis. Therefore the new basis elements can be taken as the columns of the inverse of

$$\begin{bmatrix} 1 & 1 & 0 \\ 1 & 0 & 0 \\ 1 & 0 & 1 \end{bmatrix}$$

i.e. $\{(0, 1, 0), (1, -1, -1), (0, 0, 1)\}$.

2.36 Take

$$g((x_1, x_2), (y_1, y_2)) = \tfrac{1}{2}[f((x_1, x_2), (y_1, y_2)) + f((y_1, y_2), (x_1, x_2))]$$
$$= x_1 y_1 + \tfrac{3}{2}(x_1 y_2 + x_2 y_1) + x_2 y_2$$

and

$$h((x_1, x_2), (y_1, y_2)) = \tfrac{1}{2}[f((x_1, x_2), (y_1, y_2)) - f((y_1, y_2), (x_1, x_2))]$$
$$= -\tfrac{1}{2} x_1 y_2 + \tfrac{1}{2} x_2 y_1.$$

We have $q(x_1, x_2) = f((x_1, x_2), (x_1, x_2)) = x_1^2 + 3x_1 x_2 + x_2^2$ and so the matrix of q relative to the standard basis is

$$\begin{bmatrix} 1 & \tfrac{3}{2} \\ \tfrac{3}{2} & 1 \end{bmatrix}.$$

Completing squares gives $(x_1 + \tfrac{3}{2}x_2)^2 - \tfrac{5}{4}x_2^2$. The signature is then 0 and the rank is 2. The form is neither positive definite nor positive semi-definite.

2.37 In matrix notation, the quadratic form is

$$\mathbf{x}^t A \mathbf{x} = \begin{bmatrix} x & y & z \end{bmatrix} \begin{bmatrix} 4 & -1 & 1 \\ -1 & 4 & -1 \\ 1 & -1 & 4 \end{bmatrix} \begin{bmatrix} x \\ y \\ z \end{bmatrix}.$$

It is readily seen that the eigenvalues of A are 3 (of algebraic multiplicity 2) and 6. An orthogonal matrix P such that $P^t AP$ is diagonal is

$$P = \begin{bmatrix} 1/\sqrt{6} & 1/\sqrt{2} & 1/\sqrt{3} \\ 2/\sqrt{6} & 0 & -1/\sqrt{3} \\ 1/\sqrt{6} & -1/\sqrt{2} & 1/\sqrt{3} \end{bmatrix}.$$

Changing coordinates by setting

$$\begin{bmatrix} u \\ v \\ w \end{bmatrix} = P^t \begin{bmatrix} x \\ y \\ z \end{bmatrix}$$

transforms the original quadratic form to

$$3u^2 + 3v^2 + 6w^2$$

which is positve definite.

2.38   (1) We have

$$2y^2 - z^2 + xy + xz = 2(y + \tfrac{1}{4}x)^2 - \tfrac{1}{8}x^2 + xz - z^2$$
$$= 2(y + \tfrac{1}{4}x)^2 - \tfrac{1}{8}(x - 4z)^2 + z^2.$$

Thus the rank is 3 and the signature is 1.

(2) In $2xy - xz - yz$ put $x = X + Y, y = X - Y, z = Z$ to obtain

$$2(X^2 - Y^2) - (X + Y)Z - (X - Y)Z$$
$$= 2X^2 - 2Y^2 - 2XZ$$
$$= 2(X - \tfrac{1}{2}Z)^2 - \tfrac{1}{2}Z^2 - 2Y^2.$$

Thus the rank is 3 and the signature is -1.

(3) In $yz + xz + xy + xt + yt + zt$ put

$$x = X + Y, \quad y = X - Y, \quad z = Z, \quad t = T.$$

Then we obtain

$$(X^2 - Y^2) + (X - Y)Z + (X + Y)Z + (X + Y)T + (X - Y)T + ZT$$
$$= X^2 - Y^2 + 2XZ + 2XT + ZT$$
$$= (X + Z + T)^2 - Y^2 - Z^2 - T^2 - ZT$$
$$= (X + Z + T)^2 - (T + \tfrac{1}{2}Z)^2 - \tfrac{3}{4}Z^2 - Y^2.$$

Thus the rank is 4 and the signature -2.

2.39   (1) The matrix in question is

$$A = \begin{bmatrix} 1 & -1 & 2 \\ -1 & 2 & -3 \\ 2 & -3 & 9 \end{bmatrix}.$$

Now

$$x^2 + 2y^2 + 9z^2 - 2xy + 4xz - 6yz$$
$$= (x - y + 2z)^2 + y^2 + 5z^2 - 2yz$$
$$= (x - y + 2z)^2 + (y - z)^2 + 4z^2$$
$$= \xi^2 + \eta^2 + \zeta^2$$

where $\xi = x - y + 2z, \eta = y - z, \varsigma = 2z$. Then

$$z = \tfrac{1}{2}\varsigma$$
$$y = \eta + \tfrac{1}{2}\varsigma$$
$$x = \xi + \eta - \tfrac{1}{2}\varsigma$$

so if we let

$$P = \begin{bmatrix} 1 & 1 & -\tfrac{1}{2} \\ 0 & 1 & \tfrac{1}{2} \\ 0 & 0 & \tfrac{1}{2} \end{bmatrix}$$

then we have

$$\begin{bmatrix} x \\ y \\ z \end{bmatrix} = P \begin{bmatrix} \xi \\ \eta \\ \varsigma \end{bmatrix}$$

and $P^t A P = \text{diag}\{1, 1, 1\}$.

(2) Here the matrix is

$$A = \begin{bmatrix} 0 & 2 & 0 \\ 2 & 0 & 1 \\ 0 & 1 & 0 \end{bmatrix}.$$

Now

$$
\begin{aligned}
4xy + 2yz &= (x+y)^2 - (x-y)^2 + 2yz \\
&= X^2 - Y^2 + (X - Y)z \qquad [X = x+y, Y = x-y] \\
&= (X + \tfrac{1}{2}z)^2 - Y^2 - Yz - \tfrac{1}{4}z^2 \\
&= (X + \tfrac{1}{2}z)^2 - (Y + \tfrac{1}{2}z)^2 \\
&= \xi^2 - \eta^2,
\end{aligned}
$$

where $\xi = x + y + \tfrac{1}{2}z, \eta = x - y + \tfrac{1}{2}z$ and $\varsigma = z$, say. Then

$$x = \tfrac{1}{2}(\xi + \eta - \varsigma)$$
$$y = \tfrac{1}{2}(\xi - \eta)$$
$$z = \varsigma$$

so if we let

$$P = \begin{bmatrix} \tfrac{1}{2} & \tfrac{1}{2} & -\tfrac{1}{2} \\ \tfrac{1}{2} & -\tfrac{1}{2} & 0 \\ 0 & 0 & 1 \end{bmatrix}$$

then we have

$$\begin{bmatrix} x \\ y \\ z \end{bmatrix} = P \begin{bmatrix} \xi \\ \eta \\ \varsigma \end{bmatrix}$$

and $P^t A P = \text{diag}\{1, -1, 0\}$.

(3) Here we have

$$A = \begin{bmatrix} 1 & 1 & 0 & -1 \\ 1 & 4 & 3 & -4 \\ 0 & 3 & 1 & -7 \\ -1 & -4 & -7 & -4 \end{bmatrix}.$$

The quadratic form is

$$x^2 + 4y^2 + z^2 - 4t^2 + 2xy - 2xt + 6yz - 8yt - 14zt$$
$$= (x + y - t)^2 + 3y^2 + z^2 - 5t^2 + 6yz - 6yt - 14zt$$
$$= (x + y - t)^2 + 3(y + z - t)^2 - 2z^2 - 8t^2 - 8zt$$
$$= (x + y - t)^2 + 3(y + z - t)^2 - 2(z + 2t)^2$$
$$= \xi^2 + \eta^2 - \varsigma^2,$$

where $\xi = x + y - t, \eta = \sqrt{3}(y + z - t), \varsigma = \sqrt{2}(z + 2t)$ and $\tau = t$ say. Then

$$x = \xi - \tfrac{1}{\sqrt{3}}\eta + \tfrac{1}{\sqrt{2}}\varsigma - 2\tau$$
$$y = \tfrac{1}{\sqrt{3}}\eta - \tfrac{1}{\sqrt{2}}\varsigma + 3\tau$$
$$z = \tfrac{1}{\sqrt{2}}\varsigma - 2\tau$$
$$t = \tau$$

and so

$$P = \begin{bmatrix} 1 & -1/\sqrt{3} & 1/\sqrt{2} & -2 \\ 0 & 1/\sqrt{3} & -1/\sqrt{2} & 3 \\ 0 & 0 & 1/\sqrt{2} & -2 \\ 0 & 0 & 0 & 1 \end{bmatrix}$$

gives

$$\begin{bmatrix} x \\ y \\ z \\ t \end{bmatrix} = P \begin{bmatrix} \xi \\ \eta \\ \varsigma \\ \tau \end{bmatrix}$$

and $P^t A P = \text{diag}\{1, 1, -1, 0\}$.

2.40 Here we have

$$\sum_{r<s}(x_r - x_s)^2$$

$$= (x_1 - x_n)^2 + (x_2 - x_n)^2 + \cdots + (x_{n-1} - x_n)^2$$
$$+ (x_1 - x_{n-1})^2 + (x_2 - x_{n-1})^2 + \cdots + (x_{n-2} - x_{n-1})^2$$
$$\cdots$$
$$+ (x_1 - x_2)^2$$
$$= (n-1)(x_1^2 + \cdots + x_n^2)$$
$$- 2(x_1 x_2 + \cdots + x_1 x_n + x_2 x_3 + \cdots + x_2 x_n + \cdots + x_{n-1} x_n)$$
$$= \mathbf{x}^t A \mathbf{x}$$

where

$$A = \begin{bmatrix} n-1 & -1 & -1 & \ldots & -1 \\ -1 & n-1 & -1 & \ldots & -1 \\ -1 & -1 & n-1 & \ldots & -1 \\ \vdots & \vdots & \vdots & \ddots & \vdots \\ -1 & -1 & -1 & \ldots & n-1 \end{bmatrix}.$$

Now, by adding $\frac{1}{n-1}$ times the first column to columns $2, \ldots, n$ and adding $\frac{1}{n-1}$ times the first row to rows $2, \ldots, n$, then multiplying rows $2, \ldots, n$ and columns $2, \ldots, n$ by $\sqrt{\frac{n-1}{n}}$, we see that A is congruent to the matrix

$$\begin{bmatrix} n-1 & 0 & 0 & \ldots & 0 \\ 0 & n-2 & -1 & \ldots & -1 \\ 0 & -1 & n-2 & \ldots & -1 \\ \vdots & \vdots & \vdots & \ddots & \vdots \\ 0 & -1 & -1 & \ldots & n-2 \end{bmatrix}.$$

Repeating this process we can show that A is congruent to the matrix

$$\begin{bmatrix} n-1 & 0 & 0 & 0 & \ldots & 0 \\ 0 & n-2 & 0 & 0 & \ldots & 0 \\ 0 & 0 & n-3 & -1 & \ldots & -1 \\ 0 & 0 & -1 & n-3 & \ldots & -1 \\ \vdots & \vdots & \vdots & \vdots & \ddots & \vdots \\ 0 & 0 & -1 & -1 & \ldots & n-3 \end{bmatrix}.$$

Continuing in this way, we see that A is congruent to the diagonal matrix
$$\text{diag}\{n-1, n-2, n-3, \ldots, 2, 1, 0\}.$$
Consequently the rank is $n-1$ and the signature is $n-1$.

2.41 We have
$$\sum_{r,s=1}^{n} (\lambda rs + r + s)x_r x_s$$
$$= \lambda \sum_{r,s=1}^{n} (rx_r)(sx_s) + \sum_{r,s=1}^{n} (rx_r)x_s + \sum_{r,s=1}^{n} x_r(sx_s)$$
$$= \lambda \left(\sum_{r,s=1}^{n} rx_r\right)^2 + 2\left(\sum_{r=1}^{n} x_r\right)\left(\sum_{r=1}^{n} rx_r\right)$$
$$= \lambda(x_1 + 2x_2 + \cdots + nx_n)^2$$
$$+ 2(x_1 + \cdots + x_n)(x_1 + 2x_2 + \cdots + nx_n).$$

Now let
$$y_1 = x_1 + 2x_2 + \cdots + nx_n$$
$$y_2 = x_1 + x_2 + \cdots + x_n$$
$$y_3 = x_3$$
$$\vdots$$
$$y_n = x_n.$$

Then the form is $\lambda y_1^2 + 2y_1 y_2$ which can be written as
$$\begin{cases} \lambda(y_1 + \frac{1}{\lambda}y_2)^2 - \frac{1}{\lambda}y_2^2 & \text{if } \lambda \neq 0; \\ \frac{1}{4}(y_1 + y_2)^2 - \frac{1}{4}(y_1 - y_2)^2 & \text{if } \lambda = 0. \end{cases}$$

Hence in either case the rank is 2 and the signature is 0.

2.42 Since λ is an eigenvalue of A there exist $a_1, \ldots, a_n$ not all zero such that $A\mathbf{x} = \lambda \mathbf{x}$ where $\mathbf{x} = [a_1 \ \ldots \ a_n]^t$. Then $\mathbf{x}^t A\mathbf{x} = \lambda \mathbf{x}^t \mathbf{x}$ and so, if $Q = \mathbf{x}^t A\mathbf{x}$ then we have
$$Q(a_1, \ldots, a_n) = \lambda(a_1^2 + \cdots + a_n^2).$$

2.43 Let $Q(x,x) = x^t A^t A x = (Ax)^t A x$. Since $\det A \neq 0$ we may apply the non-singular linear transformation described by $y = Ax$ so that $Q(x,x) \mapsto Q(y,y)$ where
$$Q(y,y) = y^t y = y_1^2 + \cdots + y_n^2.$$
Thus Q is positive definite.

2.44   Let $\{u_1, \ldots, u_n\}$ be an orthonormal basis of the real inner product space $\mathbb{R}^n$ under the inner product given by $\langle x|y \rangle = f(x, y)$. Let $x = \sum_{i=1}^n x_i u_i$ and $y = \sum_{i=1}^n y_i u_i$. Then

$$f(x, y) = \langle x|y \rangle = \langle \sum_{i=1}^n x_i u_i \mid \sum_{i=1}^n y_i u_i \rangle = \sum_{i=1}^n x_i y_i$$

and, for some real symmetric matrix $B = [b_{ij}]_{n \times n}$,

$$g(x, y) = \sum_{i=1}^n \sum_{j=1}^n b_{ij} x_i y_j = \mathbf{x}^t B \mathbf{y}$$

where

$$\mathbf{x} = \begin{bmatrix} x_1 \\ \vdots \\ x_n \end{bmatrix}, \qquad \mathbf{y} = \begin{bmatrix} y_1 \\ \vdots \\ y_n \end{bmatrix}.$$

Now we know that there is an orthogonal matrix P such that

$$P^t B P = \text{diag}\{\lambda_1, \ldots, \lambda_n\};$$

i.e. that there is an ordered basis $\{v_1, \ldots, v_n\}$ that is orthonormal (relative to the inner product determined by f) and consists of eigenvectors of B. Let $x = \sum_{i=1}^n \xi_i v_i$ and $y = \sum_{i=1}^n \eta_i v_i$. Then, relative to this orthonormal basis, we have

$$f(x, y) = \sum_{i=1}^n \xi_i \eta_i;$$

$$g(x, y) = \sum_{i=1}^n \lambda_i \xi_i \eta_i.$$

Consequently,

$$Q_f(x) = f(x, x) = \sum_{i=1}^n \xi_i^2;$$

$$Q_g(x) = g(x, x) = \sum_{i=1}^n \lambda_i \xi_i^2.$$

Observe now that

$$g - \lambda f \text{ degenerate} \iff (\exists x)(\forall y) \quad (g - \lambda f)(x, y) = 0$$
$$\iff (\exists x)(\forall y) \quad g(x, y) - \lambda f(x, y) = 0.$$

Since, from the above expressions for $f(x, y)$ and $g(x, y)$ computed relative to the orthonormal basis $\{v_1, \dots, v_n\}$,

$$g(x, y) - \lambda f(x, y) = \sum_{i=1}^{n} (\lambda_i - \lambda) \xi_i \eta_i$$

it follows that $g - \lambda f$ is degenerate if and only if $\lambda = \lambda_i$ for some i.

Suppose now that A, B are the matrices of f, g respectively with respect to some ordered basis of $\mathbb{R}^n$. If

$$\mathbf{x} = \begin{bmatrix} x_1 \\ \vdots \\ x_n \end{bmatrix}, \quad \mathbf{y} = \begin{bmatrix} y_1 \\ \vdots \\ y_n \end{bmatrix}$$

are the coordinate vectors of x, y relative to this basis then we have

$$g(x, y) - \lambda f(x, y) = \mathbf{x}^t (B - \lambda A) \mathbf{y}.$$

Thus we see that $g - \lambda f$ is degenerate if and only if λ is a root of the equation $\det(B - \lambda A) = 0$.

For the last part, observe that the matrices of $2xy + 2yz$ and $x^2 - y^2 + 2xz$ relative to the canonical basis of $\mathbb{R}^3$ are respectively

$$A = \begin{bmatrix} 0 & 1 & 0 \\ 1 & 0 & 1 \\ 0 & 1 & 0 \end{bmatrix}, \quad B = \begin{bmatrix} 1 & 0 & 1 \\ 0 & -1 & 0 \\ 1 & 0 & 0 \end{bmatrix}.$$

Since

$$\det(B - \lambda A) = \det \begin{bmatrix} 1 & -\lambda & 1 \\ -\lambda & -1 & -\lambda \\ 1 & -\lambda & 0 \end{bmatrix} = \lambda^2 + 1$$

the equation $\det(B - \lambda A) = 0$ has no solutions. But, as observed above, if a simultaneous reduction to sums of squares were possible, such solutions would be the coefficients in one of these sums of squares. Since neither of the given forms is positive definite, the conclusion follows.

2.45 The exponent is $-\mathbf{x}^t A \mathbf{x}$ where

$$A = \begin{bmatrix} 1 & \frac{1}{2} & \frac{1}{2} \\ \frac{1}{2} & 1 & \frac{1}{2} \\ \frac{1}{2} & \frac{1}{2} & 1 \end{bmatrix}.$$

The quadratic form $\mathbf{x}^t A\mathbf{x}$ is positive definite since

$$x^2 + y^2 + z^2 + xy + xz + yz = (x + \tfrac{1}{2}y + \tfrac{1}{2}z)^2 + \tfrac{3}{4}y^2 + \tfrac{3}{4}z^2 + \tfrac{1}{2}yz$$
$$= (x + \tfrac{1}{2}y + \tfrac{1}{2}z)^2 + \tfrac{3}{4}(y + \tfrac{1}{3}z)^2 + \tfrac{2}{3}z^2$$

which is greater than 0 for all $\mathbf{x} \neq 0$. So the integral converges to $\pi^{3/2}/\sqrt{\det A}$, i.e. to $\sqrt{2\pi^3}$.

Test paper 1

Time allowed : 3 hours
(Allocate 20 marks for each question)

1   Let V be a finite-dimensional vector space. Prove that if $f \in \mathcal{L}(V, V)$
     then
     (a) $\dim V = \dim \operatorname{Im} f + \dim \operatorname{Ker} f$;
     (b) the properties
         (i) f is surjective,
         (ii) f is injective,
         are equivalent;
     (c) $V = \operatorname{Im} f \oplus \operatorname{Ker} f$ if and only if $\operatorname{Im} f = \operatorname{Im} f^2$;
     (d) $\operatorname{Im} f = \operatorname{Ker} f$ if and only if the following properties are satisfied
         (i) $f^2 = 0$,
         (ii) $\dim V = n$ is even,
         (iii) $\dim \operatorname{Im} f = \frac{1}{2} n$.

2   Suppose that $t \in \mathcal{L}(\mathbb{C}^5, \mathbb{C}^5)$ is represented with respect to the basis

$$\{(1,0,0,0,0), (1,1,0,0,0), (1,1,1,0,0), (1,1,1,1,0), (1,1,1,1,1)\}$$

by the matrix

$$\begin{bmatrix} 1 & 8 & 6 & 4 & 1 \\ 0 & 1 & 0 & 0 & 0 \\ 0 & 1 & 2 & 1 & 0 \\ 0 & -1 & -1 & 0 & 1 \\ 0 & -5 & -4 & -3 & -2 \end{bmatrix}.$$

Find a basis of $\mathbb{C}^5$ with respect to which the matrix of t is in Jordan
normal form.

3   Let $\varphi_1, \ldots, \varphi_n \in (\mathbb{R}^n)^d$. Prove that the solution set C of the linear inequalities

$$\varphi_1(x) \geq 0, \ \varphi_1(x) \geq 0, \ \ldots, \varphi_n(x) \geq 0$$

satisfies

(a) $\alpha, \beta \in C \Longrightarrow \alpha + \beta \in C$;
(b) $\alpha \in C, t \in \mathbb{R}, t \geq 0 \Longrightarrow t\alpha \in C$.

Show that if $\varphi_1, \ldots, \varphi_n$ form a basis of $(\mathbb{R}^n)^d$ then

$$C = \{t_1\alpha_1 + \cdots + t_n\alpha_n \mid t_i \in \mathbb{R}, t_i \geq 0\}$$

where $\{\alpha_1, \ldots, \alpha_n\}$ is the basis of $\mathbb{R}^n$ dual to the basis $\{\varphi_1, \ldots, \varphi_n\}$.
Hence write down the solution of the system of inequalities

$$\varphi_1(x) \geq 0, \ \varphi_2(x) \geq 0, \ \varphi_3(x) \geq 0, \ \varphi_4(x) \geq 0$$

where $\varphi_1 = [4, 5, -2, 11], \varphi_2 = [3, 4, -2, 6], \varphi_3 = [2, 3, -1, 4]$ and $\varphi_4 = [0, 0, 0, 1]$.

4   Let A be a real orthogonal matrix. If $(A - \lambda I)^2 x = 0$ and $y = (A - \lambda I)x$ show, by considering $y^t y$, that $y = 0$. Hence prove that an orthogonal matrix satisfies an equation without repeated roots.

Prove that a real orthogonal matrix with all its eigenvalues real is necessarily symmetric.

5   Prove that if a real quadratic form in n variables is reduced by a real non-singular linear transformation to a form

$$\sum_{i=1}^{n} \lambda_i y_i^2$$

having p positive, q negative, and $n - p - q$ zero coefficients then p and q do not depend on the choice of transformation.

For the form

$$a_1 x_1 x_2 + a_2 x_2 x_3 + \cdots + a_{n-1} x_{n-1} x_n$$

in which each $a_i \neq 0$, show that $p = q$; and for the form

$$a_1 x_1 x_2 + a_2 x_2 x_3 + \cdots + a_{n-1} x_{n-1} x_n + a_n x_n x_1$$

in which each $a_i \neq 0$, show that

$$|p - q| = \begin{cases} 0 & \text{if } n \text{ is even;} \\ 1 & \text{if } n \text{ is odd.} \end{cases}$$

Test paper 2

Time allowed : 3 hours
(Allocate 20 marks for each question)

1   Let V be a finite-dimensional vector space and let $e \in \mathcal{L}(V,V)$ be a projection. Prove that

$$\operatorname{Ker} e = \operatorname{Im}(\operatorname{id}_V - e).$$

If $t \in \mathcal{L}(V,V)$ show that $\operatorname{Im} e$ is t–invariant if and only if $e \circ t \circ e = t \circ e$; and that $\operatorname{Ker} e$ is t–invariant if and only if $e \circ t \circ e = e \circ t$. Deduce that $\operatorname{Im} e$ and $\operatorname{Ker} e$ are t–invariant if and only if e and t commute.

2   If $U = [u_{rs}] \in \operatorname{Mat}_{n \times n}(\mathbb{C})$ is given by

$$u_{rs} = \begin{cases} 1 & \text{if } s = r + 1; \\ 0 & \text{otherwise,} \end{cases}$$

and $J = [j_{rs}] \in \operatorname{Mat}_{n \times n}(\mathbb{C})$ is given by

$$j_{rs} = \begin{cases} 1 & \text{if } r + s = n + 1; \\ 0 & \text{otherwise,} \end{cases}$$

show that $U^t = JUJ$. Deduce that if $A \in \operatorname{Mat}_{n \times n}(\mathbb{C})$ then there is an invertible matrix P such that $P^{-1}AP = A^t$.

Find such a matrix P when A is the matrix

$$\begin{bmatrix} 0 & 4 & 4 \\ 2 & 2 & 1 \\ -3 & -6 & -5 \end{bmatrix}.$$

3 Let V be a vector space of dimension n over a field F. Suppose that W is a subspace of V with $\dim W = m$. Show that

(a) $\dim W^\perp = n - m$;

(b) $(W^\perp)^\perp = W$.

If $f, g \in V^d$ are such that there is no $\lambda \in F \setminus \{0\}$ with $f = \lambda g$, show that $\operatorname{Ker} f \cap \operatorname{Ker} g$ is of dimension $n - 2$.

4 Let V be a finite-dimensional complex inner product space and let $f : V \to V$ be a normal transformation. Prove that

$$f^2(x) = 0 \implies f(x) = 0$$

and deduce that the minimum polynomial of f has no repeated roots.

If $e : V \to V$ is a projection, show that the following statements are equivalent :

(a) e is normal;

(b) e is self-adjoint;

(c) e is the orthogonal projection of V onto $\operatorname{Im} e$.

Show finally that a linear transformation $h : V \to V$ is normal if and only if there are complex scalars $\lambda_1, \ldots, \lambda_k$ and self-adjoint projections $e_1, \ldots, e_k$ on V such that

(1) $f = \lambda_1 e_1 + \cdots + \lambda_k e_k$;

(2) $\operatorname{id}_V = e_1 + \cdots + e_k$;

(3) $(i \neq j)$ $e_i \circ e_j = 0$.

5 (a) Show that the quadratic form $x^t A x$ is positive definite if and only if there exists a real non-singular matrix P such that $A = PP^t$. Show also that if $\sum_{i,j=1}^{n} b_{ij} x_i x_j > 0$ for all non-zero vectors x then $\sum_{i,j=1}^{n} b_{ij} p_i x_i p_j x_j \geq 0$ for all x. Hence show that if $x^t A x$ and $x^t B x$ are both positive definite then so is

$$\sum_{i,j=1}^{n} a_{ij} b_{ij} x_i x_j.$$

(b) For what values of k is the quadratic form

$$\sum_{r=1}^{n} x_r^2 + k \sum_{i<j} x_i x_j$$

positive definite?

Test paper 3

Time allowed : 3 hours
(Allocate 20 marks for each question)

1 If U, W are subspaces of a finite-dimensional vector space V prove that

$$\dim U + \dim W = \dim(U + W) + \dim(U \cap W).$$

Suppose now that $V = U \oplus W$. If S is any subspace of V prove that

$$2\dim S - \dim V \le \dim[(U \cap S) \oplus (W \cap S)] \le \dim S.$$

In the case where $V = \bigoplus_{i=1}^{n} U_i$ find similar upper and lower bounds for

$$\dim \bigoplus_{i=1}^{n} (U_i \cap S).$$

2 For the matrix

$$A = \begin{bmatrix} 0 & 1 & 0 \\ -1 & 1 & 1 \\ -1 & 0 & 2 \end{bmatrix}$$

find a non-singular matrix P such that $P^{-1}AP$ is in Jordan normal form.

If λ is positive, obtain real values of x_{ij} such that $C^2 = D$ where

$$C = \begin{bmatrix} x_{11} & x_{12} & x_{13} \\ 0 & x_{22} & x_{23} \\ 0 & 0 & x_{33} \end{bmatrix}, \quad D = \begin{bmatrix} \lambda & 1 & 0 \\ 0 & \lambda & 1 \\ 0 & 0 & \lambda \end{bmatrix}.$$

Hence, or otherwise, find a real matrix X such that $X^2 = A$.

3 Let V be a vector space of dimension n over a field F and let W, X be subspaces of V. Prove that

$$(W + X)^\perp = W^\perp \cap X^\perp \quad \text{and} \quad (W \cap X)^\perp = W^\perp + X^\perp.$$

Given $g_1, \ldots, g_n \in V^d$, prove that the following conditions concerning $f \in V^d$ are equivalent :

(1) $\bigcap_{i=1}^n \operatorname{Ker} g_i \subseteq \operatorname{Ker} f$;

(2) f is a linear combination of $g_1, \ldots, g_n$.

4 Show that the matrix

$$A = \begin{bmatrix} \frac{1}{2} & \frac{1}{2} & \frac{1}{2} & \frac{1}{2} \\ \frac{1}{2} & \frac{1}{2} & -\frac{1}{2} & -\frac{1}{2} \\ \frac{1}{2} & -\frac{1}{2} & \frac{1}{2} & -\frac{1}{2} \\ -\frac{1}{2} & \frac{1}{2} & \frac{1}{2} & -\frac{1}{2} \end{bmatrix}$$

is orthogonal. Find its eigenvalues and show that the matrix $A^2 - I_4$ has characteristic equation

$$X^2(X^2 - 3X + 3) = 0.$$

Find a unitary matrix U such that $U^{-1}AU$ is diagonal.

5 Let $Q(k, r)$ be the quadratic form

$$k(x_1^2 + x_2^2 + \cdots + x_r^2) - (x_1 + x_2 + \cdots + x_r)^2.$$

Show that

$$(k - 1)Q(k, r) = kQ(k - 1, r - 1) + y_r^2$$

where y_r is a homogeneous linear function of $x_1, \ldots, x_r$.

Hence find the rank and signature of

$$n(x_1^2 + x_2^2 + \cdots + x_n^2) - (x_1 + x_2 + \cdots + x_n)^2.$$

Test paper 4

Time allowed : 3 hours
(Allocate 20 marks for each question)

1 Let F be the vector space of infinitely differentiable complex functions and let P_n be the subspace of complex polynomial functions of degree less than n. For every $\lambda \in \mathbb{C}$ define $P_{n,\lambda} = \{e^{\lambda z}p \mid p \in P_n\}$. Show that $P_{n,\lambda}$ is a subspace of F and that

$$B = \{e^{\lambda z}\frac{z^k}{k!} \mid k = 0,\ldots,n-1\}$$

is a basis of $P_{n,\lambda}$. If D denotes the differentiation map prove that

(1) $D^n(e^{-\lambda z}f) = e^{-\lambda z}(D - \lambda\,\mathrm{id})^n f$;
(2) $P_{n,\lambda} = \mathrm{Ker}(D - \lambda\,\mathrm{id})^n$;
(3) $P_{n,\lambda}$ is D–invariant;
(4) B is a cyclic basis for $D - \lambda\,\mathrm{id}$.

If $D_{n,\lambda}$ denotes the restriction of D to $P_{n,\lambda}$ find the characteristic polynomial of $D_{n,\lambda}$. If $\mu \neq \lambda$ show, by considering the characteristic polynomial of $D_{n,\lambda} + (\mu - \lambda)\,\mathrm{id}$, that $(D_{n,\lambda} - \mu\,\mathrm{id})^n$ is invertible.

2 Given $A = \begin{bmatrix} a & b \\ c & d \end{bmatrix}$ use the Cayley–Hamilton theorem and euclidean division to show that every positive power of A can be written in the form

$$A^n = \beta_1 I_2 + \beta_2 A.$$

If the eigenvalues of A are λ_1, λ_2 show that

$$A^n = \begin{cases} \dfrac{\lambda_2\lambda_1^n - \lambda_1\lambda_2^n}{\lambda_2 - \lambda_1} I_2 + \dfrac{\lambda_2^n - \lambda_1^n}{\lambda_2 - \lambda_1} A & \text{if } \lambda_1 \neq \lambda_2; \\ (1-n)\lambda_1^n I_2 + n\lambda_1^{n-1} A & \text{if } \lambda_1 = \lambda_2. \end{cases}$$

Hence solve the system of difference equations

$$x_{n+1} = x_n + 2y_n$$
$$y_{n+1} = 2x_n + y_n$$

where $x_1 = 0$ and $y_1 = 1$.

3   Suppose that $f \in \mathcal{L}(\mathbb{C}^n, \mathbb{C}^n)$ and that every eigenvalue of f is 0. Show that f is nilpotent and explain how to find $\dim \operatorname{Ker} f$ from the Jordan normal form of f.

Let $f, g \in \mathcal{L}(\mathbb{C}^6, \mathbb{C}^6)$ be nilpotent with the same minimum polynomial and $\dim \operatorname{Ker} f = \dim \operatorname{Ker} g$. Show that f, g have the same Jordan normal form. By means of an example show that this fails in general for $f, g \in \mathcal{L}(\mathbb{C}^7, \mathbb{C}^7)$.

Deduce that if $s, t \in \mathcal{L}(\mathbb{C}^n, \mathbb{C}^n)$ have the same characteristic polynomial

$$(X - \alpha_1)^{k_1}(X - \alpha_2)^{k_2} \cdots (X - \alpha_r)^{k_r}$$

and the same minimum polynomial, and if

$$\dim \operatorname{Ker}(s - \alpha_i \operatorname{id}) = \dim \operatorname{Ker}(t - \alpha_i \operatorname{id})$$

for $1 \leq i \leq r$, then s and t have the same Jordan normal form provided $k_i \leq 6$ for $1 \leq i \leq r$.

4   Let V be a vector space of dimension n over a field F.

(i) If $s \in \mathcal{L}(V, V)$ show that $s \circ s = 0$ if and only if $\operatorname{Im} s \subseteq \operatorname{Ker} s$, in which case $\dim \operatorname{Im} s \leq \frac{1}{2}n$.

(ii) Let $p \in \mathcal{L}(V, V)$ be such that $p^n = 0$ and $p^{n-1} \neq 0$. Show that there is a basis $B = \{x_1, \ldots, x_n\}$ of V such that $p(x_j) = x_{j+1}$ for $j = 1, \ldots, n-1$ and $p(x_n) = 0$.

Show that if $t = \sum_{i=1}^{n} \lambda_i p^{i-1}$ where each $\lambda_i \in F$ then t commutes with p. Conversely, suppose that $t \in \mathcal{L}(V, V)$ commutes with p and is represented relative to the basis B by the matrix $[\alpha_{ij}]_{n \times n}$. Prove by induction that

$$(j = 1, \ldots, n) \qquad t(x_j) = \sum_{i=1}^{n-j+1} \alpha_{i1} x_{i+j-1}$$

and deduce that t is a linear combination of $\operatorname{id}, p, \ldots, p^{n-1}$.

5   Show that each of the quadratic forms

$$6x_1^2 + 5x_2^2 + 7x_3^2 - 4\sqrt{2}x_2x_3,$$
$$7y_1^2 + 6y_2^2 + 5y_3^2 + 4y_1y_2 + 4y_2y_3$$

can be reduced by an orthogonal transformation to the same form

$$a_1z_1^2 + a_2z_2^2 + a_3z_3^2.$$

Obtain an orthogonal transformation which will convert the first of the above forms into the second.